INDUSTRIAL LOCATION

MICHAEL J. WEBBER
Department of Geography
McMaster University

SAGE PUBLICATIONS
The Publishers of Professional Social Science
Beverly Hills   London   New Delhi

For information address:

SAGE Publications, Inc.
275 South Beverly Drive
Beverly Hills, California 90212

SAGE Publications India Pvt. Ltd.
M-32 Market
Greater Kailash I
New Delhi 110 048 India

SAGE Publications Ltd
28 Banner Street
London EC1Y 8QE
England

Printed in the United States of America

International Standard Book Number 0-8039-2546-8
International Standard Book Number 0-8039-2325-2 (pbk.)

Library of Congress Catalog Card No. L.C. 84-050797

SECOND PRINTING, 1985

CONTENTS

INTRODUCTION TO
THE SCIENTIFIC GEOGRAPHY SERIES

Scientific geography is one of the great traditions of contemporary geography. The scientific approach in geography, as elsewhere, involves the precise definition of variables and theoretical relationships that can be shown to be logically consistent. The theories are judged on the clarity of specification of their hypotheses and on their ability to be verified through statistical empirical analysis.

The study of scientific geography provides as much enjoyment and intellectual stimulation as does any subject in the university curriculum. Furthermore, scientific geography is also concerned with the demonstrated usefulness of the topic toward explanation, prediction, and prescription.

Although the empirical tradition in geography is centuries old, scientific geography could not mature until society came to appreciate the potential of the discipline and until computational methodology became commonplace. Today, there is widespread acceptance of computers, and people have become interested in space exploration, satellite technology, and general technological approaches to problems on our planet. With these prerequisites fulfilled, the infrastructure needed for the development of scientific geography is in place.

Scientific geography has demonstrated its capabilities in providing tools for analyzing and understanding geographic processes in both human and physical realms. It has also proven to be of interest to our sister disciplines, and is becoming increasingly recognized for its value to professionals in business and government.

The Scientific Geography Series will present the contributions of scientific geography in a unique manner. Each topic will be explained in a small book, or module. The introductory books are designed to reduce the barriers of learning; successive books at a more advanced level will follow the introductory modules to prepare the reader for contemporary developments in the field. The Scientific Geography Series begins with several important topics in human geography, followed by studies in other branches of scientific geography. The modules are intended to be used as classroom texts and as reference books for researchers and professionals. Wherever possible, the series will emphasize practical utility and include real-world examples.

We are proud of the contributions of geography, and are proud in particular of the heritage of scientific geography. All branches of geography should have the opportunity to learn from one another; in the past, however, access to the contributions and the literature of scientific geography has been very limited. I beleive that those who have contributed significant research to topics in the field are best able to bring its contributions into focus. Thus, I would like to express my appreciation to the authors for their dedication in lending both their time and expertise, knowing that the benefits will by and large accrue not to themselves but to the discipline as a whole.

—*Grant Ian Thrall*
Series Editor

SERIES EDITOR'S INTRODUCTION

In this book, Professor Michael Webber analyzes the strategy and pattern of the location of industrial production. After reviewing data sources and the history of manufacturing, Professor Webber discusses the principles that govern the location decisions of firms. It should be of particular interest to students of public policy analysis to read Webber's arguments supporting the contention that industrial location incentives and tax policies have not been significant long-term factors of industrial location; rather, Professor Webber demonstrates that as transport costs have fallen, the main location factors have become labor and agglomeration. In turn, both labor and agglomeration are themselves dependent upon the general economic, political, and social system.

Webber uses numerous data illustrations to support the theoretical arguments in this book. He concludes with three examples that illustrate his industrial location analysis: (1) the aircraft parts industry in New England; (2) the industrial decline in the United Kingdom; and (3) the location pattern of manufacturing within cities.

The stress that Professor Webber places on the historical context of decisions and on the social production of labor and agglomeration characteristics are novel issues for an introductory treatment of location theory.

—Grant Ian Thrall
Series Editor

INDUSTRIAL LOCATION

MICHAEL J. WEBBER
McMaster University

1. MOTIVATION

Location Problems

Location is a concept that means where something is in relation to other things. So *industrial location* means a statement not just of the spatial distribution of industry, but also of the relations between that distribution and other phenomena. Industrial location theory explains the spatial distribution of industry by referring to other aspects of society.

The theory of industrial location was first developed to investigate a simple fact—that industrial activity is unevenly distributed over space. Figure 1.1 shows the uneven distribution of manufacturing employment in the United States in 1977. The 50 Standard Metropolitan Statistical Areas (SMSAs) on the map are those that had the largest manufacturing employment in 1977; together, they accounted for 49 percent of all U.S. manufacturing employment. Thus, manufacturing employment is largely concentrated in only a few metropolitan areas. Furthermore, even though important places like Los Angeles, Houston, and Dallas are outside it, the "manufacturing belt" that extends from the Chicago region eastward contained most of those SMSAs and 60 percent of all U.S. manufacturing employment in 1977. So the primary problem is how to explain a map like Figure 1.1.

Yet you should not regard this figure as the "facts" of industrial location in the United States. There are many other aspects of location: spatial distribution of industries; location patterns of industry within SMSAs; distribution of industries by size of town or state. Also, Figure 1.1 is a

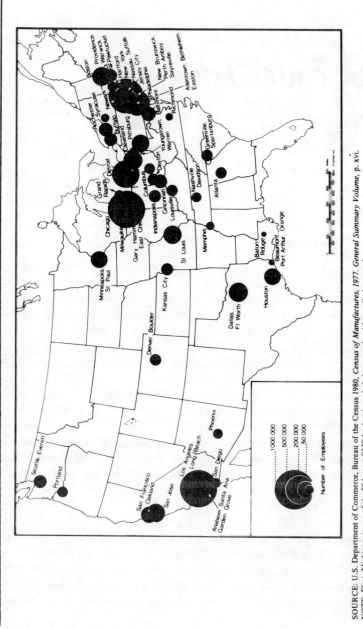

SOURCE: U.S. Department of Commerce, Bureau of the Census 1980, *Census of Manufactures, 1977. General Summary Volume*, p. xvi.

NOTE: Flint, Michigan, is one of the 50 largest SMSAs, but is excluded to comply with nondisclosure rules.

Figure 1.1 Manufacturing Employment in 50 US SMSAs, 1977

static picture, a time slice of a pattern that is changing. So there is no single location pattern, but many, each representing one issue in the location of industry.

Even though the study of industrial location began with maps like Figure 1.1, it has now gone on to study a far wider range of issues. Some of these topics are the following:

(1) Firms buy partially manufactured commodities from other firms: For example, refrigerator makers buy steel, electrical motors, packaging, and so forth. These sales of commodities from one industry to another are called *industrial linkages*. They link the performance of one industry to that of another and help explain why different industries have similar location patterns. What are the major linkages and what are their effects on location?

(2) The industrial employment in each state has gradually changed over the years. Several processes have contributed to these changes: Some factories close; some factories reduce their labor forces; some factories change their locations (relocate); some expand their employment; and new factories are set up. These various processes are the *components of change*. Different components of change have different causes. Why?

(3) It is sometimes said—and we shall examine this later—that the growth of industry in a region depends partly on the availability, cost (or wage), and degree of unionization of labor there. Yet the availability, cost, and union membership of labor itself is influenced by the growth of industry. How, then, do local *labor markets* operate?

(4) The traditional concern of industrial geography has been the production of and trade in things that can be felt and touched. Yet as corporations have grown more complex, so new phenomena have become significant: information flows, the organization of corporations, the location of research facilities and financial operations. How are these more complex organizations reflected in industrial location?

(5) If you watch any television at all, you are bombarded with advertisements for such new products as microcomputers and electronic games and by news of the effects of such new technologies as word processing. The relations between technical change, industrial change, and regional development are central to the study of industrial location: But what are these relations?

Just as this list of issues has evolved from the question of uneven development, so the theory of industrial location has broadened too. Industrial location theory was first simple and examined the location decision of firms in an abstract world. Such isolation of location from other facets of economic organization is inadequate, and now industrial location theorists try to make their theories more general, by showing how industrial structure and location reflect the broader changes that are taking place in society.

Significance of Industrial Location

These are some of the issues studied by industrial geographers. They are important for two main types of reasons.

One reason is the answer of pure science. The way in which industrial activity is distributed over space reflects the nature of our economic life; it is one aspect of the way our society works. We ought to know how our society works—it's a matter of understanding ourselves—and part of this knowledge is industrial location.

There are also reasons having to do with practical application. First, the location of industry affects people in their everyday lives: If you want to work in a particular kind of job, where do you have to live? If you live in this particular city, what types of jobs are available for you? In planning your future, you should know where jobs are and how they are changing. But also in thinking about your future, you should be aware of the kinds of changes that have affected earlier generations of workers: of the speed with which technology takes away jobs in one region of industry and adds them in another; of the way in which industries can leave seemingly secure cities for other regions or countries; of the growth and decline of different cities and regions.

Further, governments have an interest in industrial location. Industrial growth and decline, and the places of growth and decline, determine unemployment rates in different parts of the nation: Can local industrial policies reduce unemployment rates? Do welfare policies affect the location of industry? The growth of industries in a city or state affects the number of jobs there (and so the services required) as well as the governments' ability to raise taxes to pay for those local services.

Finally, the location strategies of firms influence the policies of unions. Workers' unions try to unite workers in a struggle for better pay and working conditions; firms naturally attempt to avoid places where unions are powerful. So if an industry is fixed in a location, workers have a powerful bargaining position; but if firms can easily escape places where unions are powerful, workers need to develop offsetting strategies. Many people and groups can thus learn from the study of industrial location how a changing economic environment affects the economy of local regions and how that, in turn, influences their decisions.

Plan

This book introduces the study of industrial location in the following five chapters. Chapter 2 sets the context for the study. It describes industries, data sources, types of business organizations, and changing economic characteristics. This chapter claims that the location of industry reflects the economic character of society. Chapter 3 describes the ways in which firms make location decisions: It states the principles of location theory. Chapters 4 and 5 follow the statement of principle by developing a simple theory of industrial location (least cost theory) and by showing how the various costs of production vary over space. These chapters devise some general rules about the location of industry. Chapter 6 concludes by showing how to apply these rules to the more complex contexts described in Chapter 2 in order to explain actual location patterns.

Further Reading

Another readable introduction to the study of industry is E. C. Estall and R. O. Buchanan's *Industrial Activity and Economic Geography* (1973, pp. 15-24). Doreen Massey and Richard Meegan, in their *The Anatomy of Job Loss* (1982, pp. 3-13), raise many of the questions of industrial location that arise from the recent industrial decline of Britain.

2. CONTEXT

This chapter defines the limits to the study of industrial location. It has two parts. First, the objects of study of industrial location are defined. Chapter 1 introduced the issues tackled in the location of industry; that introduction is now sharpened by distinguishing branches of economic activity and types of economic organizations. Second, the location of industry can only be understood by referring to other aspects of our economic life, so this part of the chapter describes some of the ways in which the organization of economic life has been changing: How does economic life now differ from that of the mid-nineteenth century (when modern industry began to develop in the United States) and how do those differences affect the spatial organization of society? One of the central claims of this book is that the location of industry at any time can be understood in terms of two things—the way in which economic life is organized and the principle of location. This chapter, then, tackles the first of these issues by providing the context within which the location of industry must be studied.

Objects of Study

What are the objects that industrial geographers study? The previous chapter used terms like *industry* and *firm* without defining terms: What kinds of industry? What kinds of firms? This question is now tackled in four parts: First, work is defined and methods of measuring it are described; second, manufacturing is defined; third, industries are distinguished; and finally, the types of organization that operate industries are described.

WORK

Industry is an aspect of *work*. In practice, government statisticians define work in terms of a market system. If I perform a service or make a commodity, and I am paid by an employer or a customer, then I am said to work. (Payment may be in kind rather than money—for example, sales people receive cars and restaurant workers receive meals as part of their pay.) This concept of work excludes many hard and necessary jobs: Growing fruit or vegetables in your garden for your own private consumption is excluded; cleaning your own house is excluded (but included if you pay a housekeeper); cooking for your family is excluded; rearing children is excluded. However, regular unpaid work in a family business is included. The economic model that underlies this definition is that of an exchange economy: Work occurs if labor is exchanged for money. Statistically, then, work is a narrow concept.

Many sources provide data about the amount and location of work. Data are obtained either directly from people themselves (e.g., by population census or labor force survey) or from employers (e.g., economic census or *County Business Patterns* [U.S. Department of Commerce]). Each document uses slightly different definitions and, drawing on different sources, includes different categories of workers. The main American sources are now briefly described (Canadian sources are listed in Table 2.1).

The *Census of Population* is the most massive compilation of data about the population. The Bureau of the Census (U.S. Department of Commerce) conducts such a census every ten years; most recently (to date) one was taken on April 1, 1980. Data are published for many areas: states, metropolitan areas, state economic areas, counties, incorporated places, and even parts of cities. People were counted as employed if they were at least 16 years of age and in the week preceding the census either (1) did some work as paid employees (or in their own business or farm) or (2) had a

15

TABLE 2.1   Sources of Canadian Data on the Geography of Work

(1)   Statistics Canada. Various years. *Census of Population*. Ottawa: Author.
Gives occupational and industrial structure by place of residence and place of work. No nondisclosure problems, but no data about firms' operations.

(2)   Statistics Canada. Annual. *Manufacturing Industries of Canada: National and Provincial Areas* (catalogue 31-203). Ottawa: Author.
For Canada and its provinces, provides data for each of the 20 SIC industries and for as many of the three- and four-digit industries as are compatible with nondisclosure rules on number of males and females employed (average of four month's employment), wages or salaries paid; separated into production workers (for whom total hours worked are also presented) and administrative, office, and other nonmanufacturing employees. Data from sample of manufacturing establishments and some other data are estimated.

(3)   The above reference provides data for all industries with a two- to three-year lag (the 1979 data were published in 1982). For some industries, data are published earlier in individual industry reports.

(4)   Statistics Canada. Annual. *Manufacturing Industries of Canada: Sub-Provincial Areas* (catalogue 31-209). Ottawa: Author.
Data are the same as for source 2, but are presented for census divisions (such as counties), economic regions, Census Metropolitan Areas, and other smaller cities. Nondisclosure rules limit the industrial breakdown.

job but were ill, on strike, or on vacation. Volunteers and houseworkers were specifically excluded. Unemployed people were those not employed but looking and available for work. The sum of the employed and unemployed is the civilian labor force. The census also asked how many weeks each person worked in 1979.

The Bureau of the Census in the U.S. Department of Commerce publishes *County Business Patterns* annually. These data are obtained from first quarter taxation returns filed by establishments. (An establishment is a place where business is conducted or industrial operations are performed.) An establishment's employment is the number of paid employees in the pay period including March 12, but the number of hours worked in that pay period is not counted. Establishments are classified by county.

The third main source of data about industries is the *Census of Manufactures*. This census is taken every five years by the Bureau of the Census, U.S. Department of Commerce, and is supplemented by annual sample surveys of manufacturers. Like *County Business Patterns,* these data are obtained from businesses and so are subject to nondisclosure rules: Because of confidentiality, employment is reported only for the larger counties and cities: Thus good data for smaller places are hard to obtain. Unlike *County Business Patterns*, however, the *Census of Manufactures* collects data from all establishments, not just those that employ more than four people; the *Census of Manufactures* also lists a greater variety of data than does *County Business Patterns*. . The employment data in the *Census of Manufactures* are averages of the number of employees in each of four months; they are supplemented by data on the number of hours employed during the year. The *Census of Manufactures* is the basic source for the statistics presented in this book.

Some of the differences between these three sources are instructive. In all three, people are counted as employed if they find some paid work. Hence fluctuations in the amount of work that are caused by variations in hours worked per week are concealed. (The *Census of Manufactures* does, however, publish hours worked for the year.) The *Census of Population* reports people by place of work and place of residence, but the *Census of Manufactures* and *County Business Patterns* report only place of work. *County Business Patterns* includes only workers in private businesses, thus excluding government workers, those who work in domestic service for pay, and self-employed persons. The *Census of Population* provides the finest areal breakdown because it is unaffected by nondisclosure rules.

Statistical definitions of employment, then, are based on the economic model of an exchange economy. The statistical concept of employment is the notion of labor for reward rather than the idea of work. By and large, published data do not measure well the fluctuations in hours worked. Some sources include data for only some establishments. These characteristics affect measurements of employment in different regions. For example, if one person had a full-time job but was replaced by two part-timers, then the measured level of employment increased. Similarly, employment may appear to be lower in regions where more people are self-employed than were people mainly work for others. As government publications are the only feasible source of employment data, they must therefore be used with some care.

MANUFACTURING

There are many different types of employment. The *Census of Population* classifies employment in two ways: by the nature of the job (accountant or programmer or welder) and by the nature of the industry (steelworking or education or automobile manufacturer). The *Census of Manufactures* and *County Business Patterns* use only the latter classification. From the point of view of the worker, the nature of the job is important, but location theory has largely ignored that in favor of the industrial classification, which is more important for government and business. This book largely follows that practice, although it does discuss the difference between direct production and management.

Location theorists first distinguish activities that produce (i.e., make things) from those that consume and from those that facilitate production or consumption. Production activities include wheat farming, commercial fishing, iron ore mining, operating a hamburger stand, staging an opera, construction, and transport (which changes a thing at one place into a thing somewhere else). In production, labor and machinery transform or assemble input materials to produce an output that is socially more useful than were the inputs; they directly increase the material well-being of society. Consumption activities are the purchase and enjoyment of products. Facilitating activities enable production or consumption to take place; they include retailing and wholesaling, defense and police work (which controls the physical space of production), government (which organizes the social context of production), and doctors and teachers who keep us healthy and educated. (In practice these distinctions blur: A hamburger store makes hamburgers—production—but it also sells them to customers—a facilitating activity.)

Location theorists divide production into three categories. The first class is farming, fishing, forestry, and mining, which are *extractive* activities producing primary raw materials. Labor and machinery organize and harvest the production of nature. Extractive activities depend on natural conditions; their location is studied in land use and resource theory. The second class includes activities that are sold directly to consumers at their point of production, including industries such as restaurants and theaters, transport, and construction. People must travel to such "factories" as restaurants and theaters to consume their output, so they locate in the same way as schools or shops, to which consumers must also travel: The location of this group is studied by central place theory (see King 1984). Transport and construction is regarded as entirely demand driven, and so ignored. The third category of production is manufacturing strictly defined; it is the object of industrial location theory. Manufacturers process raw materials and assemble semifinished goods into final products, usually in factories. Statistically, manufacturing also includes the buying, maintenance, shipping, management, engineering, and security operations of factories. (The "Introduction" to the *Census of Manufactures, Summary Volume* clearly defines manufacturing.)

The location of industry, then, is quite different from that of work. First, unpaid labor and the production of not-for-sale goods are excluded, thereby leaving statistical employment. Facilitating activities are then omitted, leaving direct production. Third, extraction and consumer-oriented activities are excluded because of their specific locational requirements. What's left is called manufacturing. Figure 1.1 is a map of manufacturing employment in the United States: It is important not to confuse this with a map of productive activities for as we have noted, manufacturing excludes a lot of production. Figure 2.1 is a comparable map of industrial employment in Canada; it shows how Canadian industry is locationally an offshoot of the American manufacturing belt.

The degree of spatial concentration of manufacturing employment in North America is high. In 1977, the manufacturing belt of states from Wisconsin and Illinois eastward through Michigan, Indiana, Ohio, and Pennsylvania to New York, New Jersey, Connecticut, and Massachusetts accounted for 49 percent of all U.S. manufacturing employment. The dominance of this belt is declining however: It contained 57 percent of the employment in 1963. California is a major outlier (8.9 percent of manufacturing employment in 1977, 8.2 percent in 1963); Texas accounts for 4.5 percent of the employment (3.0 percent in 1963); and the Southern states of Virginia, Tennessee, South Carolina, and Georgia contain 10.9 percent of the U.S. manufacturing employment (9.0 percent in 1963). The Canadian offshoot of the manufacturing belt is the Windsor to Quebec City axis, which in 1970 accounted for 72 percent of all manufacturing employment compared to 55 percent of the population in Canada (Yeates 1975, p. 27).

Employment is an (admittedly imperfect) measure of the human effort involved in production. Government statistics also describe the value of the outputs produced by the effort of labor (see *Census of Manufactures, 1972*, vol. 3, part 2, appendix A). One such measure is the *value of shipments*: The net market price received for output, excluding freight and taxes. This value includes only goods produced and sold. Now, consider Figure 2.2, which shows two ways in which a steel can may be produced. In the upper diagram, a factory buys raw materials, makes steel, and then makes the steel into cans. The value of shipments is $1 million per week. In the lower diagram, production is separated into two factories. The first buys raw material and makes steel: The value of shipments is $750,000 per week. The second factory buys steel and makes it into cans valued at $1 million per week. In the second case, then, the separation of production into different factories causes the value of shipments to rise to $1.75 million. This duplication is avoided by the concept of *value added by manufacture*. This measure is obtained by subtracting the total cost of all input materials from the value of shipments. In the upper diagram, with materials costing $600,000, the factory's value added is $400,000 ($1,000,000 value of shipments less $600,000 cost of materials). In the lower diagram, the first factory adds value equal to $150,000 (= $750,000 − $600,000) and the

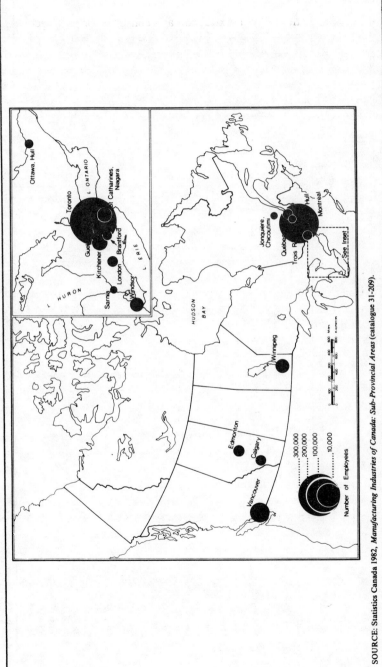

SOURCE: Statistics Canada 1982, *Manufacturing Industries of Canada: Sub-Provincial Areas* (catalogue 31-209).

Figure 2.1 Manufacturing Employment in Canadian Cities, 1978

19

TABLE 2.2 Characteristics of 25 Largest SMSAs, United States, 1977 and 1972

SMSA	1977 Rank (unit)	1977 Employment (thousand people)	1977 Value Added (billion dollars)	1977 Value Added per Employee (dollars)	1972 Rank (unit)	1972 Employment (thousand people)	1972 Value Added (billion dollars)	1972 Value Added per Employee (dollars)
Chicago	1	884	27.5	31109	1	910	21.8	23920
Los Angeles-Long Beach	2	826	24.7	29903	2	779	15.2	19560
New York	3	797	19.9	24969	3	949	15.1	15930
Detroit	4	565	18.1	32035	4	552	11.7	21201
Philadelphia	5	452	13.4	29646	5	496	9.2	18539
Houston	6	210	9.8	46667	11	163	4.2	25688
St. Louis	7	249	8.3	33333	8	256	5.2	20145
Cleveland	8	265	8.1	30566	7	269	5.2	19416
Newark	9	256	7.9	30859	6	272	5.6	20635
Dallas-Fort Worth	10	270	7.9	29259	13	230	4.1	17720
Boston	11	268	7.5	27958	9	267	4.9	18457
San Francisco-Oakland	12	193	6.8	35233	14	185	3.8	20606
Rochester	13	144	6.7	46528	10	142	4.4	30876
Pittsburgh	14	240	6.5	27083	12	263	4.2	15844
Milwaukee	15	204	6.4	31373	16	200	3.7	18500
Minneapolis-St. Paul	17	181	6.2	34254	22	134	2.9	21377
Cincinnati	18	160	5.6	35000	17	158	3.6	22596
Anaheim-Santa Ana-Garden Grove	19	178	5.3	29775	23	131	2.7	20873
Baltimore	20	166	5.2	31325	18	180	3.5	19301
Buffalo	21	140	4.7	33571	19	152	3.2	20740
Louisville	22	107	4.4	41121	20	113	3.0	26656
Norfolk-Suffolk	23	156	4.4	28205	25	(N/A)	(N/A)	
Atlanta	24	129	4.2	32558	27	132	2.5	18673
Seattle-Everett	25	132	4.2	31818	28	109	2.2	20577

SOURCE: U.S. Department of Commerce, Bureau of the Census 1980, *Census of Manufactures, 1977: General Summary*, p. xvi; and *Census of Manufactures, 1972*, vol. III, part I, p. xvii.

second factory adds value equal to $250,000 (= $1,000,000 - $750,000). The total value added is the same in each diagram. In a sense, then, value added measures the market price of work done in the factory.

Table 2.2 presents some measures of industrial activity in 1972 and 1977 for 25 American metropolitan areas (those that had the greatest value added in manufacturing in 1977). Two particular aspects of these data are important.

First, the rates of growth of employment varied widely. At the one extreme, Houston, San Jose, and Anaheim recorded percentage increases in employment of 30 percent or more; while on the other hand, employment fell in New York (by 16 percent), Philadelphia, Baltimore, and Buffalo (all by 8 percent). These differences distinguish the cities of the old manufacturing belt from the cities of the South and West.

Second, although the value added in manufacturing is generally greatest in metropolitan areas that have the largest employment, nevertheless there are differences in value added per employee. For example, in Houston, the value added per employe in 1977 was $46,667 whereas in New York it was only $24,969. The value of the work done per employee could vary for two reasons: For equivalent products, Houston's factories may receive a higher price than do New York's; or for equivalent effort per worker, Houston's factories may produce more commodities than do New York's. The latter is likely to be a major source. (Why, for example, would you pay more for a shirt made in Houston than for a shirt made in New York?)

But why do some places produce more per worker than do others? In general because in those places there is more plant and machinery per worker than in the others. Now differences between metropolitan areas in the amount of machinery per worker have two sources: Either Houston (for example) contains many industries that are highly mechanized (such as oil refining) and New York has less mechanized industries (such as clothes making), or within each industry Houston's factories are more mechanized than New York's.

Suppose the latter is the more important reason of the two. Then in each industry, metropolises like Houston, Rochester, and Louisville would contain more mechanized factories (i.e., more efficient factories) than would places like New York, Philadelphia, Boston, and Pittsburgh. But more efficient factories should be able to surpass less efficient ones and so grow faster; hence metropolitan areas such as Houston, Rochester, and Louisville should grow faster than the others. However, Figure 2.3 shows that this did not happen: Places that had a high value added per employee in 1972 did not grow faster than the others. This graph indicates that our first assump-

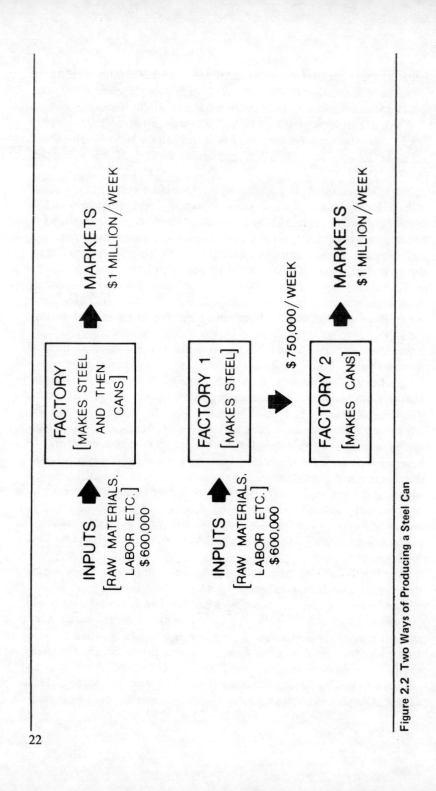

Figure 2.2 Two Ways of Producing a Steel Can

tion is wrong, and therefore that metropolitan areas in which the value added per employee is high are those in which employment is concentrated in mechanized industries. Of course, this evidence is not conclusive (direct data are needed to establish this purported fact), but it does raise an interesting question: Why do places have these different employment patterns?

INDUSTRIES

The idea that some types of manufacturing are more highly mechanized than others introduces the concept of an *industry*. We talk of oil refining or clothing as distinct industries but what is meant by an industry?

Microeconomic theory provides one theoretical basis for classifying activities into industries: the concept of the *market* for a commodity. The market for, say, wheat consists of all those people and firms who have wheat to sell and all those who wish to buy wheat. The problem is to determine the price of wheat, the amount that is sold, and the profit that producers of wheat make. If the wheat offered by every seller is identical, then consumers prefer to buy from that seller who charges the lowest price—price is the only basis for preference. So, if a commodity is homogeneous, sellers must compete for sales in terms of price. (Conversely, by persuading consumers that their commodities are not homogeneous, sellers can try to avoid price competition.) In microeconomic theory, then, a commodity is a homogeneous product, and different sellers are distinguished only by price. Provided that there are enough sellers, no consumer will buy from a seller who tries to sell at a price above the others: All sellers of the commodity therefore offer the same price. Thus arises the idea of a competitive market: Many independent sellers of a homogeneous commodity, all selling at the same price. Such a group of sellers is a theoretical industry.

In practice, however, this notion of an industry is far too limited. Some commodities are virtually homogeneous: A pharmaceutical drug, for example, is made according to a precise formula. But most commodities are like automobiles, dishwashers, and cookies in that there are many different kinds of them. Other industries are defined not by the commodity they make but by the type of buyer to whom they sell: The agricultural machinery industry sells a variety of machines to farmers. Some industries are defined by the material they process: Petrochemical industries make many products from oil. Furthermore, a single market may be supplied by two or more industries: The dairy industry and the vegetable oil industry compete to sell spreads for bread.

Whatever the theoretical issues, in practice industries are defined by government statisticians who have produced a Standard Industrial Classification. The principles of the U.S. classification are discussed in the "Introduc-

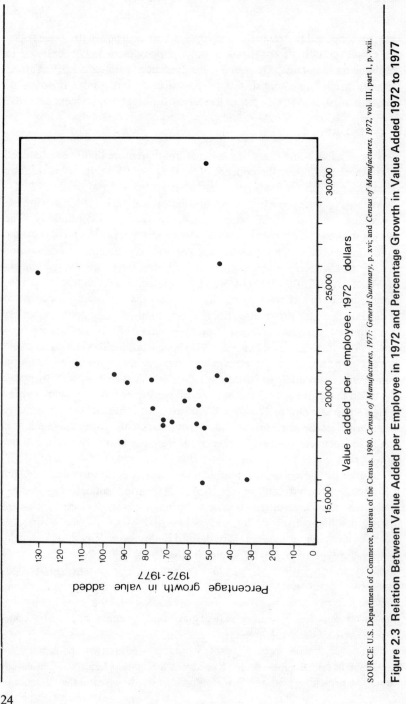

Value added per employee, 1972 dollars

Percentage growth in value added 1972-1977

SOURCE: U.S. Department of Commerce, Bureau of the Census. 1980. *Census of Manufactures, 1977: General Summary*, p. xvi; and *Census of Manufactures, 1972*, vol. III, part 1, p. vxii.

Figure 2.3 Relation Between Value Added per Employee in 1972 and Percentage Growth in Value Added 1972 to 1977

tion'' to each *Census of Manufactures*; for Canada, see Statistics Canada, *Standard Industrial Classification Manual*. The United States and Canada both define twenty *major groups*, such as food and beverage industries or paper and allied industries. Each major group is divided into three-digit industries: 143 in the United States and 112 in Canada. The United States also has a four-digit classifications comprising 451 industries. Table 2.3 illustrates the way in which this classification system works.

In the Standard Industrial Classification most industries are defined in terms of specific groups of products, usually made by similar materials and processes. But some industries are defined by the process used (aluminum foundries) and some by the material used (wood office furniture). However, no matter how careful or detailed the classification, the factories grouped in an industry do not account for all the production of that industry, and those factories also produce commodities belonging to other industries. Both these problems are caused by the fact that factories generally produce a variety of commodities. Therefore, aluminum foundries in New York do not necessarily produce the same products as aluminum foundries in Texas. Industries are not as homogeneous as we commonly think (or as we consider them for empirical research).

ORGANIZATIONS

Activity has so far been classified according to the kind of commodity it produces or service it performs. But capitalist economic sytems, such as those of North America and Northern Europe, contain many types of organizations. These organizations are responsible for different activities and have different roles to play in the economy.

Perhaps the most ignored organization in economic life is the household. The primary economic role of households now is to reproduce the labor force—that is, to satisfy the material and emotional needs of workers and to raise children who will work in the future. The inputs to this process are consumer goods. But households are also important producing units: Meals have to be produced out of food; houses and clothes have to be maintained; and gardens must produce. These two roles, production and reproduction, are not undertaken for profit (indeed, they are largely performed by unpaid labor), which distinguishes the household from other producing units in capitalist economies.

Over the last two hundred years, the importance of the household sector in production has diminished. Gardens feed fewer people now. More food is bought in processed or semiprocessed form and more machinery is used within the household. Fewer clothes are made within the home than in earlier

TABLE 2.3 Structure Of U.S. Standard Industrial Classification

Major Group	Three-Digit Industry		Four-Digit Industry	
20 Food and kindred products				
21 Tobacco manufacturers				
22 Textile mill products				
23 Apparel and other textile products				
24 Lumber and wood products				
25 Furniture and fixtures				
26 Paper and allied products	281 Industrial inorganic chemicals			
27 Printing and publishing	282 Plastics materials and synthetics			
28 Chemicals and allied products	283 Drugs		2831 Biological products	
29 Petroleum and coal products	284 Soap, detergents and toiletries		2832 Medicinal chemicals	
30 Rubber and miscellaneous plastic products	285 Paints, varnishes and allied products		2834 Pharmaceutical preparations	
31 Leather and leather products	286 Industrial organic chemicals			
32 Stone, clay and glass products	287 Agricultural chemicals			
33 Primary metal industries	289 Miscellaneous chemical products			
34 Fabricated metal products				
35 Machinery, except electrical				
36 Electrical and electronic equipment				
37 Transportation equipment				
38 Instruments and related products				
39 Miscellaneous manufacturing				

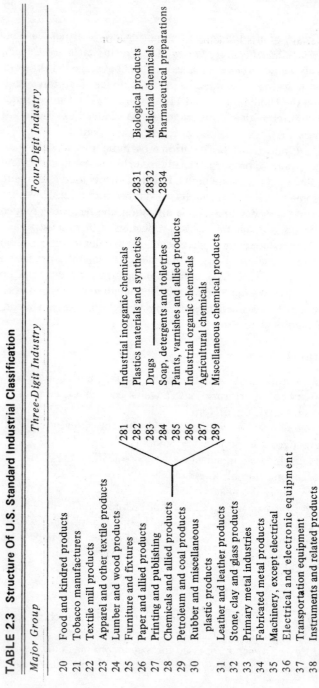

SOURCES: U.S. Department of Commerce, Bureau of the Census, 1976. *Census of Manufactures, 1972*, vol. 3, part 2, appendix. For Canada, see Statistics Canada, 1971. *Standard Industrial Manual, Revised 1970* (catalogue 12-501).

times. Less entertainment is homemade. In fact, the process whereby the commercial sector (which produces for profit) has replaced the household (not-for-profit) sector as a producer of consumer goods and has mechanized the remaining household production is one of the major changes that have taken place in economic life during this century. Thus, more and more *goods* (things of use to people) are being produced as *commodities* (goods made by the commercial sector for sale).

The commercial sector is made up of *firms*. Industrial business is owned by firms, which operate plants and employ labor to produce output. Several types of business are recognized legally. The *proprietorship* is a business owned by one person: The owner organizes everything and in return reaps all the profit (or loss). The *partnership* is also privately owned, by two or more owners who cooperate in the venture. *Cooperatives* are organizations of producers (or consumers) who share both the task of organizing the business and the profits. But by far the most important type of business organization, in terms of its control over the economy, is the *corporation* (called a limited company in Britain), which brings together the money capital of many large and small shareholders who jointly own a single organization. Unlike private companies, corporations are typically run by professional, hired managers rather than by their owners.

Firms operate establishments. An *establishment* is a place at which business is done, a physically distinct unit (not necessarily a legally distinct unit). Examples of establishments are a factory (or plant), a warehouse, a research laboratory, a sales office, and a company headquarters. So locationally the simplest kind of firm has only one establishment—factory, warehouse, and headquarters combined—and makes a single product. Here, such an establishment is called simply a factory and the organization a factory-firm. Generally, however, *firm* here means a business organization—either independently owned or a division of a larger operation—that itself owns several establishments and runs them as a single, interrelated business. The locationally (and legally) most complex type of organization is here called the *corporation*, which is a multiproduct and multiestablishment organization comprising several firms.

A factory-firm has a single establishment that is the site of all its activities. Under one roof are performed the actual manufacturing, the direct management of the factory, the research and development of new products, the sales effort, and the long-run planning. However as firms and, more particularly, corporations have evolved, these various activities have often been separated into different establishments. Now manufacturing can be divided not only into separate industries but also into its separate functions—

production and day-to-day management (factory), research and development (laboratory), and sales and planning (headquarters).

These different functions of business exhibit distinct location patterns. Table 2.4 presents some evidence of this for the United States. The main fact that this table illustrates is that the states of the old manufacturing belt have a higher proportion of America's manufacturing company managers than they do of its direct operators, whereas the manufacturing states of the South and West—California, Georgia, North Carolina, Texas, and Virginia—all contain a smaller proportion of the managers than they do of the operatives. In fact in 1977, nearly 55 percent of the managerial workers in manufacturing were employed in Illinois, Michigan, New Jersey, New York, Ohio, and Pennsylvania. This proportion has fallen since 1963, mainly because of the precipitous decline of New York state as an employer of manufacturing managers and operatives.

Evolution of Manufacturing

Industries and the organizations that run them have changed dramatically during the last two centuries. These changes have caused the location of industry to change too, for the location of industry reflects the broader structure of society. This section describes two aspects of the evolution of manufacturing since the mid-nineteenth century: changes in the organization of production and changes in the types of production. It is important to study the manner in which manufacturing has evolved, first because the present contains relict features of the past (New England mill towns, the industrial northeast, inner industrial districts of the old manufacturing cities), and second because the relative importance of the various factors that affect location has changed over time.

In the mid-nineteenth century, firms and plants were small. Table 2.5 shows some of the relevant data for the United States. In 1869, the average manufacturing establishment employed eight people and used only 1.14 horsepower of machinery per worker. Most firms were unincorporated, either proprietorships or partnerships. Also in the nineteenth century, wage labor was much less prevalent than it is today (Table 2.6): In 1870 less than half of the population was in the labor force; even by the turn of the century, wages and salaries of employees accounted for only 55 percent of national income. Conversely, there were many manufacturing establishments—one for every 111 people in 1870.

The manufacturing sector was a capitalist sector. It was organized by owners to produce commodities for sale using wage labor, with the goal of making profits. Many activities are not organized in this way: The household and family business do not use wage labor, nor do legal and

TABLE 2.4   Percentage Distribution of Operatives and of Managerial Workers, Selected U.S. States, 1963 and 1977

State	Percentage of U.S. Managers in State		Percentage of U.S. Operatives in State		Ratio, 1977, of Managerial Percentage: Operative Percentage
	1963	1977	1963	1977	
California	6.5	5.5	8.3	9.1	0.60
Connecticut	2.1	3.2	2.5	2.0	1.60
Delaware	2.4	2.3	0.3	0.2	11.50
Georgia	0.6	1.4	2.2	2.5	0.56
Illinois	8.3	8.9	7.1	6.4	1.39
Indiana	1.4	1.6	3.7	3.7	0.43
Massachusetts	3.8	3.1	4.0	3.1	1.00
Michigan	11.4	9.8	5.4	5.5	1.78
Minnesota	2.1	2.8	1.4	1.6	1.75
Missouri	2.6	2.7	2.3	2.2	1.22
New Jersey	7.1	8.1	4.8	3.7	2.19
New York	20.0	11.7	10.5	7.5	1.56
North Carolina	1.7	2.4	3.2	4.0	0.60
Ohio	7.5	7.5	7.3	6.8	1.10
Pennsylvania	9.5	8.1	8.2	6.7	1.21
Tennessee	0.6	0.8	2.0	2.6	0.31
Texas	2.1	4.1	3.1	4.5	0.91
Virginia	1.0	1.0	1.8	2.1	0.48
Wisconsin	1.9	2.1	2.8	2.8	0.75

SOURCE: U.S. Department of Commerce, Bureau of the Census 1980. *Census of Manufactures, 1977: General Summary,* pp. 1-12 to 1-14.
NOTE: States are only included if they contained 2% of U.S. operatives or managers in 1977. Observe the effects of state taxation provisions for incorporation on the managerial/operatives ratio for Delaware.

medical services. Agriculture, when run by family farms, produces some commodities not for sale and uses little wage labor. In Table 2.6, the sum of employee compensation and corporate profits shows that at the turn of the century some 62 percent of the U.S. national income arose in capitalist organizations (remember that the national income does omit not-for-sale production).

Within manufacturing, firms seek profits. There are three main avenues whereby profits can be made. First, a firm can seek to enlarge its profits by producing its commodities more efficiently than do rival firms, thus making a greater profit per unit of output or, if it reduces its prices, enlarging its share of the market. This is the route of technical change in the production of existing commodities. Second, a firm can seek to produce new

TABLE 2.5 Selected Aspects of U.S. Manufacturing Industry, 1870-1980

| Year | Number of Establishments (thousands) | Number of Employees | | Value Added (billions of dollars) | Horsepower per Worker (unit) |
		Production (thousands)	Other (thousands)		
1977	360	13700	5900	585	–
1967	311	13955	4537	262	–
1954	287	12372	3272	117	9.58
1947	241	11900	2400	81	–
1929	207	8370	1290	31	4.91
1899[a]	205	4502	348	4.65	2.18
1899[b]	509	5098	380	5.48	–
1869[b]	252		2054	1.40	1.14

SOURCES: U.S. Department of Commerce, Bureau of the Census. 1975. *Historical Statistics of the United States, Colonial Times to 1970*, pp. 666 and 681; and *Statistical Abstract of the United States, 1981*, p. 780.
a. Factories excluding hand and neighborhood industries.
b. Factories including hand and neighborhood industries.

TABLE 2.6 U.S. Labor Force and National Income Characteristics, 1900-1980

| | Labor Force as Percentage of Population | | Percentage of National Income Accruing to: | | | |
Year	All	Nonagricultural, Civilian, Employed	Employees	Business[a]	Rent[b]	Profits[c]	Interest[d]
1977	62.8	54.9	–	–	–	–	–
1967	60.6	52.9	71.1	10.0	3.3	12.4	3.2
1954	57.6	46.5	–	–	–	–	–
1947	56.8	45.7	65.5	15.6	3.8	14.1	0.9
1929	56.2	42.2	63.0	15.8	6.6	6.4	8.1
1900	53.7	–	55.0	23.6	9.1	6.8	5.5

SOURCES: U.S. Department of Commerce, Bureau of the Census. 1975. *Historical Statistics of the United States, Colonial Times to 1970*, pp. 127 and 236.

NOTE: Where data are missing, they are unavailable.

a. Income of unincorporated business.
b. Rental income of persons.
c. Corporate profits before tax.
d. Net interest.

31

commodities—televisions as well as radios, cars as well as bicycles, or microcomputers as well as scientific measuring devices. This is the route of new product development. Associated with this route is the third method of raising profit: enlarging the sphere of capitalist commodity production by taking over activities previously performed by noncapitalist organizations. In particular, this method has taken over much household production: food preparation, house maintenance, and clothing production have all been increasingly performed outside the household and mechanized. All three routes to profit have affected the scope, type, and organization of production.

Table 2.6 contains some indexes of the effect of enlarging the sphere of capitalist commodity production. The nonagricultural civilian labor force has more than doubled as a proportion of the population, and the labor force is now 60 percent of the population: There has been a decline in the number of people left at home to work in the household. Wages or salaries and profits account for over 80 percent of the U.S. national income (compared to 62 percent in 1900), so property and private business income now comprise only one-eighth of the total of all income. Thus, an increasing proportion of the population works outside the home (must travel to work) and works for pay in profit-seeking firms.

New product development and increasing participation of the household sector together imply that the range of commodities produced by manufacturers is now significantly larger than it was before. One obvious effect of this process has been the creation of entirely new industries: the chemical, instruments, and electrical and electronic machinery industries were all insignificant in the nineteenth century. Manufacturing now means a wider set of processes than it used to. Two other aspects of this process deserve further comment: the product cycle, and research and development.

It has been claimed that new products are introduced into the market in a characteristic pattern. When an innovation is produced and a new commodity is first made, firms are uncertain about the design, production method, and market for the commodity. The market is small and uncertain, so firms are small (or small parts of corporations) and labor intensive. As the market for the commodity is enlarged by advertising and by its acceptance as a part of daily life, and as firms acquire appropriate production techniques, profitability rises. These profits encourage firms to seek larger market shares by standardizing the commodity and mechanizing its production. Prices now fall and output rises until everyone is using the commodity; as this expansion phase ends, the only sales are those due to the replacement of obsolete items and the increase in consumption per head. This history, with its phases of initial production, high profits, standardization, and market saturation, is called the *product cycle* hypothesis. It claims that at different phases, manufacturing establishments differ and so may have different locational needs.

Table 2.7 illustrates one aspect of the product cycle hypothesis—the rates of growth of U.S. industry groups at different times. Some industries were well developed during the nineteenth century and have never grown faster than has manufacturing as a whole since 1900: textile mill products (SIC 22), apparel (23), lumber and wood products (24), and leather industries (31, apart from the aberration of 1919). These industries were in the saturated phase before 1900. Another group reached their peak growth rates before 1919—tobacco (21), printing and publishing (27), petroleum and coal products (29), and primary metals (33). The third group comprises industries that grew fastest since World War II: chemicals (28) and instruments (38) in the 1950s; paper and paper products (26) and machinery (35, 36) in the 1960s; furniture (25), rubber and plastics (30), and stone, clay, and glass products (32) in the 1970s. The transport industry (37) has two peaks—one in the period between 1899 and 1919 and one in the 1950s. Such product cycles are even more clearly seen in the data for three- and four-digit industries.

The path to greater profits via new products has spawned an entire twentieth century industry: research and development. However, it is the first route to profits—technical change—that has had the main effect on the organization of production. Cost-reducing technical changes within factories have taken the form of new equipment, standardized production methods, and economies in the use of raw materials. Now each worker has ten times the horsepower available to nineteenth-century workers (Table 2.5), and indexes of output per hour of work show the same rise. By 1977, each establishment had 54 employees, and over one-quarter of the manufacturing workers were employed in plants where over one thousand people worked. The sheer increase in the size of factories and in the amount of machinery per worker are the major changes to have affected manufacturing work.

Increases in plant size have been accompanied by the concentration of output and capital in a few corporations in each industry. The production of many industries is now dominated by a few firms. Whereas in 1899, unincorporated businesses still accounted for 35 percent of the value added in manufacturing, this proportion had fallen to less than 5 percent by the 1960s. By this time, too, one-quarter of all the value added in manufacturing was produced by the 50 largest companies. The price of this corporate market power has been an enlargement in the proportion of nonproduction employees, from 7 percent in 1899 to over 30 percent now (see Table 2.5).

TABLE 2.7   Index Numbers of U.S. Manufacturing Production by Industry 1900-1980

	Major Group	1979	1970	1960	1947	1929	1919	1899
						Sample Years		
	All manufacturing[b]	154	105	65	39	22	13	6
20	Food[a]	148	112	78	56	26	18	11
21	Tobacco products	118	100	90	67	37	25	11
22	Textile mill products	145	112	69	55	38	26	15
23	Apparel	134	101	82	58			
24	Lumber and wood products	137	106	74	59	51	40	42
25	Furniture and fixtures	162	108	72	45			
26	Paper	151	113	68	39	20	11	4
27	Printing and publishing	137	104	70	43	31	17	5
28	Chemicals	212	120	53	18	6	3	1
29	Petroleum and coal products	144	113	77	42	23	9	2
30	Rubber and plastics	239	116	54	25	14	7	–
31	Leather and products	72	91	99	94	74	133	47
32	Stone, clay and glass	164	106	81	48	–	–	–
33	Primary metals	121	107	74	65	42	26	9
34	Fabricated metals	149	102	71	50	–	–	–
35	Nonelectrical machinery	154	104	57	–	–	–	–
36	Electric machinery	175	108	52	–	–	–	–
37	Transport equipment	135	90	64	31	20	12	2
38	Instruments	175	111	58	25	–	–	–
39	Miscellaneous	154	111	71	44	–	–	–

SOURCES: U.S. Department of Commerce, Bureau of the Census. 1975. *Historical Statistics of the United States. Colonial Times to 1970*, pp. 667-8; and *Statistical Abstract of the United States, 1981*, p. 778.
NOTE: The index numbers combine three not strictly comparable series—for the 1899 to 1947 and 1947 to 1970 from *Histrocial Statistics*, and for 1970 to 1980 from *Statistical Abstract*.
a. Short names only; full names are in Table 2.3.
b. 1967 = 100.

These three routes to greater profit (technical change, new commodities, and enlargement of the sphere of commodity production) and their effects upon the economy have been set within the context of a system in which the *state* plays a larger role now than it did formerly. (The state is the set of all government institutions, local and national.) The state provides major markets for manufacturers, particularly the armaments and space industries: By the 1960s, U.S. government purchases accounted for almost one-quarter of the output of the transport equipment and ordnance industries and one-sixth of the electrical equipment produced (U.S. Department of

Commerce, *Historical Statistics*, p. 272). The state has also supported ailing industries by running them directly (as in the United Kingdom and Canada), by guaranteeing loans, and by reducing taxes on corporate profits (in the United States from 45 percent in 1960 to 36-37 percent in 1979 and 1980; U.S. Department of Commerce, *Statistical Abstract*, p. 777). There is some claim that local variations in taxes affect industrial location, but this will be examined in Chapter 4. Particularly in the United Kingdom and Canada, the state has involved itself in regional planning (ostensibly intended to direct manufacturing plants to locations where unemployment rates are high, but it is likely that these policies have had limited effects on the location of industry).

These changes in the organization and type of production have altered the lives of all of us. They affect us both as consumers and as workers. Part of this effect occurs by changes in the location of industry, and the task of industrial location theory is to understand that link between social and locational change. To develop that understanding, we must now begin the task of theorizing location. In Chapter 6, we return to this link by applying the location principles to actual societies.

Further Reading

Some aspects of work and employment are discussed by E. B. Philips and R. T. LeGates in *City Lights* (1981, pp. 449-472). A more prosaic introduction to the notion of industrial systems and the components of those systems is contained in F. E. I. Hamilton and G. J. R. Linge (eds.), *Spatial Analysis, Industry and the Industrial Environment: 1, Industrial Systems* (1979, pp. 1-23).

3. PROFITS AND LOCATION

The previous chapters have claimed as a guiding principle that the task of location theory is to link the changes that are taking place in society (the type and organization of production) to the spatial distribution of industry. We have now reviewed the nature of the units that operate in society and have described some important changes that have occurred over the last one hundred years. The link between these changes and location is location theory. That theory has two parts: first, an understanding of the way the economy works, and then an application of that understanding to the historical context to analyze industrial location patterns. This chapter attempts the first part: to understand the way in which the economy works.

Assume what has been taken for granted—that the economies of Western Europe and North America (and particularly their manufacturing sectors) are capitalist. Production is organized by the owners of firms; the owners' goal is to sell the firm's output in order to make profits. Firms use capital to buy raw materials, semifinished goods, the plant and machinery, and to hire labor in order to produce commodities for sale. The coordination of all this activity (aggregate output and price) is achieved by the market rather than by a planner.

But what are the rules of the operation of such an economy? This question can be answered in two ways. One is to make claims about the overall goals of the economic system, such as asserting that the economy is organized to achieve maximum rates of economic growth. The other way is followed here: to analyze the economy at the level of the individual firm and to explain location patterns by referring to the decisions made by firms. This chapter explains the basis for these decisions, by examining location decisions theoretically.

Location in Theory

There are three kinds of location decisions. The first is the decision to build or buy a new establishment. The firm is starting in business or relocating its business or building additional capacity. Location theory has traditionally concentrated on this type of decision, as Chapter 4 will reveal. Second, the firm can reorganize production, by altering the products produced at its various establishments or by closing some factories and concentrating production at others. The problem then is not to add capacity but to rearrange it or even to reduce it. The third decision is the decision to close down an establishment—to reduce capacity. Often, but not always, this decision is involuntary, being forced by bankruptcy.

In each of these cases, the firm is making a decision to invest or to disinvest. This decision depends on the existence of perceived opportunities to sell (or the lack of opportunities) and on the firm's perceived ability to find the investment. Unforced location decisions are thus investment decisions, and the choice of a location depends on the perception of an opportunity.

Once established, a business must either grow or stagnate. There are business people who are happy to control small, old-fashioned firms, but they account for very little manufacturing output. Successful firms are those that survive their initial growth pains and then are able to increase their capital and scale of operations. The recent history of the high-technology computer industry of "silicon valley" (around San Jose, California) illustrates this process.

A firm can increase its scale of production only by capital investment. There are two sources of this capital. One is the investment of the firm's own profits; in 1980, firms retained over 60 percent of their after tax profits (U.S. Department of Commerce, *Statistical Abstract*, p. 777) and these retained earnings comprised over 80 percent of gross investment in equipment and buildings (p. 781). Second, capital may be obtained from investors or borrowed from financial institutions. People and institutions invest and lend in order to obtain a return on their investment, and the surest evidence about that return is a past level of profits. The more profit a firm makes, then, the more it can grow. (However, for the economy as a whole, more profit does not equal more growth because profits can be invested abroad.) In traditional economic theory, this thesis is summarized in the principle that firms seek to maximize profits—in particular, seek locations that maximize profits.

Maximizing profits means that the firm chooses the most profitable location or set of locations in view. This principle is not meant in the precise sense. Some decisions are highly constrained by the firm's circumstances so that there are only a few alternatives: For example, there may be only two or three blocks of land for sale in a given city to which the roads are good enough for heavy trucks, for which waste disposal provisions have been made, and utilities have been connected. For other decisions, the firm has greater freedom to choose. The theoretical claim is simple: Decisions are best understood as the decisions of profit-maximizing firms.

There have been two main objections to this principle. The first claims that people do not know enough to be able to choose optimum locations. The second objection states that corporations are not now run by owners but by managers, who have no personal stake in corporate profits. Let's consider the second objection first, for it is the less serious.

MANAGERIALISM

The idea of the maximum profit criterion was first popularized by Adam Smith, an eighteenth-century British economist who claimed that capitalism works by appealing to people's self-interest. It is in each person's interest to increase his or her wealth. At the dawn of industrial capitalism in the early nineteenth century, most manufacturing businesses were unincorporated firms, owned by one or a few individuals (who typically also managed the business). Owners had a direct, personal incentive to maximize the profit of firms.

Virtually all manufacturing output is now produced by large corporations. These corporations are owned by stockholders, pension funds, insurance companies, and banks; but they are run by professional managers.

The wealth of managers is not directly related to the profitability of a corporation; therefore, managers have less incentive than do owner-operators to make profit-maximizing decisions. Alternative decision criteria include security or size; that is, decisions are made in order to guarantee the future of the firm or to maximize its sales. This claim is made by those who advance behavioral theories of the firm (see, for example, Cyert and Marsh 1963, and Baumol 1959) and it is bolstered by interviews with business people who have sometimes chosen a location for what they say are personal reasons.

Despite its popularity, the behavioral argument fails. The flaw of the behavioral approach is that it personalizes the profit motive. Smith claimed that firms seek profits to satisfy the personal greed of their owners; and undoubtedly, many people do want to become rich, just as many owners obtain psychic rewards from operating their own businesses. Still the large corporations that dominate manufacturing maximize profits—not because of the personal whim of their managers, but because they have no alternative. If a firm does not keep up in the race for profits, it stagnates and in the long run, dies. Low profits imply both reduced dividends and lower retained earnings that reduce stockholders' incomes and the capital appreciation of the stock, both of which drive away investors. As the retained earnings fall, so reinvestment and product development are curtailed, which drive profits even lower when techniques and product lines become obsolete. The goal of maximum profits is not a personal whim but a structural necessity.

UNCERTAINTY

The second argument against the claim that firms make decisions to maximize profits is far more serious than the first. It is the argument that firms simply do not have the information they require to maximize profits.

Firms are uncertain: They clearly do not possess enough information to guarantee that a decision is a profit-maximizing one. A firm faces three main sources of uncertainty: Some data may be too expensive to obtain; the future can never be predicted with perfect accuracy; and other firms may make later decisions (to locate, to develop new products, or to market products) that affect the anticipated benefits. For all these reasons, the actual decision may not turn out to be as good as the firm had expected.

The problem of uncertainty is particularly acute for new location decisions. The impact of uncertainty depends on the amount of capital committed by the decision and the ease with which the decision can be revoked. The decision to hire a particular worker, Jones, commits little capital (Jones's wage) and is easily changed (Jones can be fired). The decision to produce at a particular output level for the next month is both costlier and less easily changed. Automobile manufacturers invest even more in producing a new style of car: After years of design and testing (capital investment), the choice of a design is not easily changed in less than several years. (Of course, advertising reduces the risk of failure in this case.) The location decision is more expensive still: It involves setting up or relocating the entire plant, equipment, and work force of an establishment. This expense is greater for larger establishments, but involves a smaller proportion of the total assets of multiplant corporations than those of factory firms. Also, the location decision is not easily undone—establishments cannot simply be picked up and moved elsewhere. So a poor locational choice has long-term effects on profitability.

Uncertainty affects location decisions because they are both costly and fixed. While these characteristics encourage firms to seek actively for maximum profit locations, nevertheless there will always be some uncertainty. Therefore the expected outcome of a decision must be distinguished from the actual outcome. Technically, this distinction is between ex ante and ex post evaluations.

Location decisions are made in this manner: For each of the feasible sites, find out how much can be sold, how much it costs to produce that output, and the expected prices. After obtaining these data for each site, the locations can be compared (this is the ex ante evaluation) and the most profitable one selected (this location is the ex ante optimum). When the establishment is built and running, the firm discovers the actual costs and sales at each place: This is the ex post optimum. Clearly, a firm can only seek the ex ante optimum.

This idea has been taken one step further. Investment information has two characteristics: First, it is not free, so that managers have to pay for the data on which to base decisions; and second, people have only a limited capacity to process information (although this capacity has increased as computers have become more widely used). These characteristics have prompted the claims that maximum profit locations (even ex ante optimum) are fundamentally unobtainable, and thus managers should choose some levels of profits that would satisfy them and then make decisions so as to obtain that satisfactory profit. Such decision making is called *satisficing* (attaining a satisfactory profit) as opposed to *optimizing* (obtaining the highest level of profit).

Now, a remarkable fact has been discovered that relates satisfactory and optimal decisions. Suppose that you have to choose a location for an investment. You evaluate sites one by one and compute the profit you can make at each. After each calculation you decide to choose that site or to examine the next. Any rejection you make must be final. There is a cost to evaluate each site; although this cost fluctuates, you nevertheless expect it to be constant on average. Now, what rule do you use to choose a location? It turns out that the optimal rule is to satisfice. That is, if you want to choose the location which yields you the most profit net of the cost searching, you must first decide upon a satisfactory level of profit and then choose the first location you find that yields at least that level. Satisficing is optimal!

Apparently, the simple rule that decisions are made to maximize profits needs to be clarified. Firms can only find ex ante optima, not ex post optima (except by chance); and the selection of a best site must recognize the costs of searching. It follows that even though firms are optimizing, the economy need not be located in a profit-maximizing pattern: After the event, better locations could be found than the places actually chosen. Thus we can imagine a location pattern that comprises many establishments located by firms seeking optimal (i.e., satisficing) ex ante locations, yet this location pattern could be rearranged to give every firm higher profits.

But this idea must not be overemphasized. The pattern of location in a society results not merely from firms' decisions, but also from society's evaluation of those decisions. If at a given location a firm cannot make the normal (i.e., satisfactory) rate of profit, then it cannot attract investors nor create the capital needed to renovate and expand: It will fall even further behind its peers in the industry. The attainment of a normal rate of profit is a socially imposed goal—and a firm's ex ante optimum may not be good enough to attain it. Factory closures are a constant reminder of this fact.

The process whereby society imposes its profitability criterion upon the decisions of individuals can be seen not only in plant closings but also in settlement patterns. Figure 3.1 illustrates the process. During the process of European settlement of America, many towns and villages were set up at apparently advantageous sites. There were far too many for the needs of the economy, so only the locationally fittest places survived (This argument is extended by Alchian 1950.)

These are some of the arguments that surround the claim that firms maximize profits. This claim is a sound basis for location theory despite the contrary arguments. Location decisions are ex ante optima when search costs are considered; thus a firm that has perfect information may be able to find a better site. However, the actual location pattern is a consequence not only of firms' decisions but also of the weeding out of inefficient firms by competition: The end result looks as if firms did find the profit maximum.

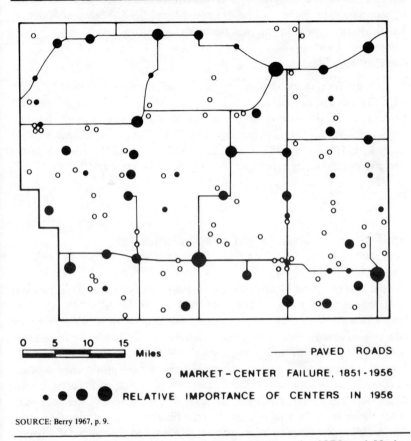

0   5   10   15   Miles   ——— PAVED ROADS

o MARKET - CENTER FAILURE, 1851 - 1956

• ● ● ● RELATIVE IMPORTANCE OF CENTERS IN 1956

SOURCE: Berry 1967, p. 9.

Figure 3.1 Relative Importance of Market Centers in 1956 and Market
Center Failure 1851-1956 in Iowa

Theory and Practice

Firms, then, seek to maximize profits, a decision rule that is strictly enforced by the dynamics of capitalist competition. The remainder of this book examines how this principle works itself out in different kinds of economic circumstance. Historically the earliest and conceptually the simplest theory investigates the location decision for a single factory. This theory is examined in Chapters 4 and 5, which augment the more abstract argument by examining the spatial variation in the factors that influence profitability. Then the theory is applied to analyses of concrete historical circumstance (Chapter 6). The theory is first presented simply and then in more detail, but is always based on the maximum profit principle.

Further Reading

An introduction to the ideas of theory and model is provided by P. E. Lloyd and P. Dicken in *Location in Space* (1977, pp. 8-17). Some general criticisms of industrial location theory—and suggestions for its improvement—are found in the papers by David Smith and Doreen Massey in F. E. I. Hamilton and G. J. R. Linge (eds.) *Spatial Analysis, Industry and the Industrial Environment: 1, Industrial Systems,* (1979, pp. 37-55, 57-72).

4. LEAST COST THEORY: TRANSPORT COSTS

The simplest, and historically the earliest, theory of industrial location is the *minimum cost theory*, first published in 1909 by Alfred Weber. The major part of Weber's theory examines the location of a factory-firm or of a single, new factory belonging to a corporation. Chapter 3 claimed that a firm makes profit-maximizing decisions: That is, it tries to collect the correct (in terms of its costs and benefits) amount of information in order to choose that location among those examined for which profits are greatest. Least cost theory simplifies this analysis by assuming that the necessary information is freely available and that the firm examines all possible sites. The problem of collecting information is thus avoided. In contrast to Chapter 3, this chapter and the next examine the location strategy of a single factory for which all relevant information is freely available. Together these two chapters present an elementary but also updated version of Weber's theory.

This chapter is in four parts. The first section describes the essential idea of least cost theory. The following section is empirical: It examines the actual structure of transport costs and rates. The theory is then advanced in the two succeeding sections, which analyze the effects of transport costs on location—the first assumes a given plant size while the second examines the interrelation between plant size and transport costs. Chapter 5 completes the theory by discussing spatial variations in production costs and the theory of agglomeration.

The Idea of Least Cost

A factory buys raw materials and semi-processed products, organizes a labor force and machinery to make a product and delivers that product to warehouses. Thus, the costs that the factory has to incur are the costs of raw materials and semi-finished goods; the costs of plant and equipment and processing in the factory; and the costs of delivering the finished product to the buyer (warehouse). On the positive side, the firm obtains a monthly revenue equal to the price at which each unit of the good is sold (say that is p) multiplied by the number of units of the product sold per month (Q, say). So the firm has a revenue $R = pQ$, and, its profit from the operation of this factory in each month is profit = revenue - costs, or

$$\pi = R - C = pQ - C \qquad [1]$$

Let's now consider each of the three components of costs—inputs, processing and delivery—in turn. The firm needs a regular supply of raw materials and semifinished goods. These purchases can be made in two ways. Some firms buy their inputs from a supplier and then make their own transport arrangement: The supplier charges a price and the buyer pays for transport. (This is buying *free-on-board*, or fob.) Other firms buy the inputs delivered to the factory: The supplier then charges a price made up of the cost of making the input plus the cost of transporting it to the factory (plus a markup for profit). (This is buying with *cost-insurance-freight*, or cif.) Either way, the cost input is in effect made up of two components: A price charged at the point of production of the input (e.g., at the mine) plus a cost of transport from the source to the buyer.

The cost of delivering outputs also has two components—the cost of making the good plus the cost of transporting it to the buyer. The final price charged by the factory thus includes a delivery cost—the cost of transporting the good to the market.

The third component of the cost is the cost of processing. This includes many elements: the monthly share of the cost of building the factory; the monthly share of the cost of machinery and of upkeep; the cost of managing the operations of the factory and of selling the output; the wages paid to workers; and the cost of running the factory—electricity, gas, water, property and other taxes and the like. All these are lumped together as processing costs.

The location decision thus has several components. The firm has to decide on the markets which the factory is to serve (and its output levels), must

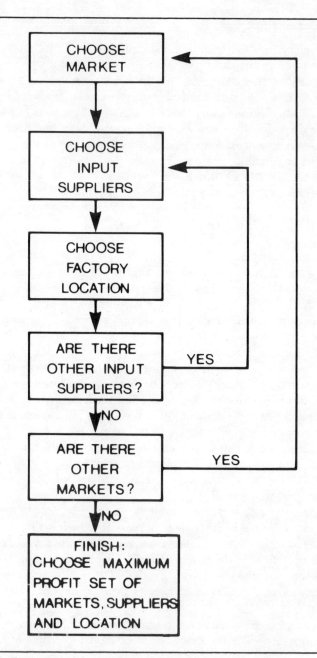

Figure 4.1 Flow Chart of a Firm's Location Decision for a Factory

arrange for suppliers of inputs, and then must choose a location for its factory. The theoretical analysis of these decisions proceeds in the way shown by Figure 4.1. First the firm must decide on an appropriate market and on the amount that it will try to sell to that market. Next it must decide what sellers of inputs are to be patronized. Given these two decisions, the firm must then find the maximum profit location. Now are there any other sources that could supply materials to the factory? If so, suppose that one of these other sources is patronized, and reconsider the maximum profit location. Continue to search for alternative sources of supply and reevaluate the location until all other suppliers have been considered. Could the factory supply any additional or different markets? If so, choose one of these markets and return to step 2. Continue iterating 5, 2, 3, and 4 until all alternative markets have been evaluated. This procedure will provide the firm with a maximum profit combination of markets (and outputs), suppliers, and location.

Suppose that step 1 has been accomplished and that a given market has been chosen. Then p and Q are known. There remains a pair of decisions—suppliers and location. Equation 1 shows that for a given pair p and Q, the choice of suppliers and of location maximizes profits if and only if it minimizes costs. That is, for given market and price decisions, the firm's choice of suppliers and locations must be a minimum cost one. The first stage in the theory of industrial location must then be to find minimum cost locations and suppliers to serve a given market—hence the name least cost theory.

So the first problem is, given a certain market, find that set of suppliers and that location that minimize the cost of serving that market (the market may comprise one city, several cities or the nation). For example, suppose that a firm wants to sell 600 chairs a month to the federal government in Washington, D.C. Each chair weighs an average of 13 pounds. The firm needs 6,000 pounds of wood per month, available from suppliers X_1, X_2, or X_3; it needs 4,000 pounds of steel tube, from Y_1 or Y_2; and it needs 1,000 pounds of miscellaneous fittings available from Z. It has six combinations of suppliers: X_1 for wood and Y_1 for tube; or X_1 for wood and Y_2 for tube; X_2 and Y_1; X_2 and Y_2; X_3 and Y_1; or X_3 and Y_2. For each combination of suppliers, a least cost location has to be found, and then the least cost combination of suppliers.

Thus the problem has become even simpler: For a given set of markets and suppliers, find the location that minimizes costs. Figure 4.2 is a map showing the location of the various suppliers and the single market, M. The figure also shows the costs of transporting the supplies and chairs as

TABLE 4.1   Calculation of Minimum Cost Location for Data in Figure 4.2

		S_1	S_2	S_3	S_4
(1)	Costs of production (in dollars)	12000	12000	12000	13000
(2)	Transport costs on fittings	400	350	300	250
(3)	Transport costs on wood from X_1	2000	0	2000	3000
(4)	Transport costs on tube from Y_2	1100	700	500	600
(5)	Transport costs on chairs to M		2500	3800	4000
	Total costs (1+2+3+4+5)	15500	16550	19600	20850
(6)	Transport costs on wood from X_1	2000	0	2000	3000
(7)	Transport costs on tube from Y_1	1200	800	600	200
	Total costs (1+2+5+6+7)	15600	16650	19700	20450
(8)	Transport costs on wood from X_2	5000	2800	2000	500
(9)	Transport costs on tube from Y_2	1100	700	500	600
	Total costs (1+2+5+8+9)	18500	18730	18600	18350
(10)	Transport costs on wood from X_2	5000	2800	2000	500
(11)	Transport costs on tube from Y_1	1200	800	600	800
	Total costs (1+2+5+10+11)	18600	18850	18700	17950
(12)	Transport costs on wood from X_3	7000	5000	4200	3100
(13)	Transport costs on tube from Y_2	1100	700	500	600
	Total costs (1+2+5+12+13)	20500	21500	23800	20950
(14)	Transport costs on wood from X_3	7000	5000	4200	3100
(15)	Transport costs on tube from Y_1	1200	800	600	200
	Total costs (1+2+5+14+15)	20600	21650	23900	20550

well as the possible sites of production S_1, S_2, S_3, and S_4, which are all near places where there is available labor and power. Table 4.1 lays out the calculations of costs if fittings came from Z, tube from Y_2, and wood from X_2. You can see that the least cost site is S_4. The table also lays out the costs of locating at S_3 for the other five combinations of suppliers. You should now complete these calculations for sites S_2, S_3 and S_4. The result is that you can find the least cost combination of suppliers and location to serve the market at M; namely, use X_1, Y_2, Z, and locate at S_1 (the market). It is a straightforward matter to do the same calculations for other market and so to find the maximum profit market, set of suppliers, and location.

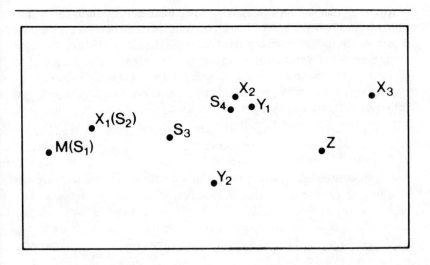

Production costs for 600 chairs per month for factory at

S_1	S_2	S_3	S_4
$12000	$13000	$12000	$13000

Transport costs on 600 lbs of wood

from to	S_1	S_2	S_3	S_4
X_1	$2000	0	$2000	$3000
X_2	$5000	$2800	$2000	$500
X_3	$7000	$5000	$4200	$3100

Transport costs on 4000 lbs of steel tube

from to	S_1	S_2	S_3	S_4
Y_1	$1200	$800	$600	$200
Y_2	$1100	$700	$500	$600

Transport costs on 100 lbs of fittings from Z to

S_1	S_2	S_3	S_4
$400	$350	$300	$250

Transport costs on 600 chairs to market, M

from	S_1	S_2	S_3	S_4
	$0	$2500	$3800	$4000

Figure 4.2 Example of a Location Problem (with data)

Evidently this sort of calculation can be made for any problem, and in practice this is what firms do if they have some freedom of choice in locating a factory. But location theory must do more than this: It must also state some general rules that allow us to interpret the economic landscape. The first part of the construction of these general rules is attempted in the following sections by ignoring production costs entirely and examining only the influence of transport costs on location. To do this, the nature of transport costs in reality must be described.

Transport Rates

Location decisions imply a particular spatial pattern of trade in commodities. Once a location decision has been made, then patterns of purchases and of market supplies are fixed. Thus, location decisions are the economic logic underlying *spatial interaction* (for an introduction to this field, see Haynes and Fotheringham 1984). Conversely, the location decision is itself influenced by the costs of transporting commodities over space. The next section examines the nature of those costs: first, by presenting some theory of transport rates, and second by providing evidence about the actual costs of transport.

THEORY

The structure of transport rates is extremely complex. This structure is now examined by considering the nature of a transport company and its costs (rates being set as costs plus a markup for profits). Suppose, then, that you buy a quantity of input and you ship it to your factory. What costs of the transport company do you have to pay for?

First, you have to pay a share of the office, sales, and organization costs of the transport company. Secretaries, warehouses, managers, traffic organizers, and the office must be paid for. These costs are proportional to the number of separate journeys that have to be organized, so there is a particular fixed cost to organize your shipment. If you are a regular user of the transport company—say you make the same shipment once a week—then the company can schedule your shipment regularly and so can discount the fixed cost for your regular business.

The fixed cost of organizing your shipment also depends on the size of your shipment in relation to the size of transporter the company uses. If your commodity fills a truck, the company simply has to organize that truck. But if you have only a small quantity, the company has to organize several shipments to use the one truck fully. In trucking, the fixed cost of organization is therefore lower for truckload lots than for less-than-truckload lots. By contrast, rail freight is almost always of less-than-trainload lots, so a rail company has to organize many shipments to put a train of freight together.

The second item you have to pay for is the loading of the truck or rail car. The truck must go to your supplier, or the rail car to a siding, and must load up. Now, the fixed cost of organization is independent on both the amount you want shipped (apart from the economy of truckload lots) and the distance of the journey; on the other hand, the *terminal cost* of loading depends partly on the amount you ship but is independent of the length of the journey. The terminal cost also depends on the fragility of the goods: A front-end loader can quickly and cheaply load ten tons of ore, but furniture must be packaged by hand and loaded more carefully.

Then you must pay for the journey itself. This cost has two components. The first is your share of the cost of the vehicle's upkeep and its right of way—maintenance, costs of insurance, road taxes in the case of a truck, and in addition rail companies must pay to operate and to maintain their track. These costs depend largely on the length of the journey and on the proportion of the total load that is yours (i.e., the volume of your load). Second, there are out-of-pocket costs attributable to the journey—the wages, fuel, and road tolls. Wages are proportional to the time taken—that is, distance. Considering that a driver and brakeman can operate a freight train of several hundred cars, the cost of wages per ton is far lower on trains than on trucks. Fuel and tolls depend on the distance of the shipment, but also increase as the weight of the load rises.

The final cost that you have to pay is unloading. This is a terminal cost, similar in structure to that at the outset of the journey.

In addition to these costs, the transport company charges you a sum to cover its profits. In practice, this profit is a standard percentage markup over costs. Thus the total charge you must pay to the transport company is total costs plus, say, 20 percent of those costs.

These arguments about the nature of transport costs are summarized in Figures 4.3 and 4.4. We have argued that rail freight is subject to greater terminal charges but lower *line haul* costs than is truck freight, and Figure 4.4 reflects this, indicating that railways should be favored for long hauls and trucks for shorter. (There is some evidence that this is the case.) One other aspect of this diagram is interesting. Consider Table 4.2 which classifies the possible ways in which the terminal and line haul costs of

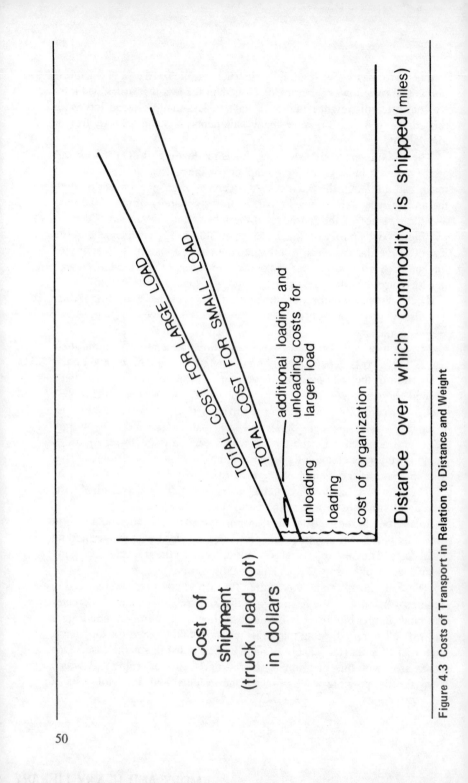

Figure 4.3 Costs of Transport in Relation to Distance and Weight

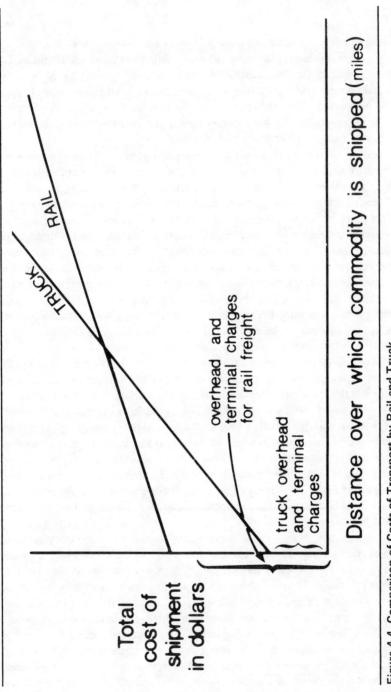

Figure 4.4 Comparison of Costs of Transport by Rail and Truck

RAIL

TRUCK

overhead and
terminal charges
for rail freight

truck overhead
and terminal
charges

Total
cost of
shipment
in dollars

Distance over which commodity is shipped (miles)

51

rail and truck could be compared. If trucks had higher terminal charges and higher line haul costs, then railways would be cheaper for all distances; why would anyone use a truck? Equally, if railways had the higher terminal charges and higher line haul costs, trucks would always be cheaper and no one would use railways. Hence the fact that both railways and trucks exist in competition with each other means that the mode having the higher terminal cost also has the lower line haul cost.

Before examining the structure of actual freight rates, some complicating factors must be introduced into this simple picture. The complications are return journeys, dead time, and practical freight rate setting.

Each outward journey must be balanced in some way by a return journey. (If not, the origin would eventually run out of vehicles and the destination would be overrun with empty trucks or cars.) If the transport company can readily find a load for the return journey, then that load can pay to return the vehicle to its origin. But if no return load exists (nor even a load for a roundabout return), then you are liable to be charged for the return journey of the vehicle as well as the outward one. In practice, this means that transporting a commodity over a route in which the volume of shipments is largely one way is more expensive than shipping a commodity over a route in which the shipments in each direction largely balance.

The out-of-pocket costs of a journey, directly attributable to you, include the wage of the driver and other workers while the vehicle is moving. But the transport company must pay the workers for a whole day, say for eight hours. So if your journey involves one hour of loading, five hours of driving, and then another hour of unloading, the terminal costs must include the two hours during which the driver waited for the truck to be loaded, and the five hours of driving must also be charged at such a rate as to pay for the eighth hour of the driver's day, when he or she can do nothing useful. Transport rates for journeys that do not take exactly a whole number of days must be calculated to pay for this dead time. (Vehicle dead time must also be paid.)

In principle, when you ask the transport company to ship your materials, an accountant should calculate the direct and indirect costs of the order. In practice the company has previously computed the direct and indirect costs of various kinds of shipment, depending on whether it is regular or not, a carload or not, its ease of loading and cartage, its weight or volume, and the distance it must be transported. But this cost will not increase linearly with each mile as shown in Figure 4.3, for practical rates are quoted in zones. For example, the following rates may be quoted: $0.40 per thousand pounds per mile for destinations within 50 miles, $0.35 per thousand pounds per mile for destinations between 50 and 120 miles and $0.30 per

TABLE 4.2 Possible Structures of Terminal and Line Haul Costs for Truck and Rail Freight

Line Haul Costs	Terminal Costs
truck > rail	truck > rail
truck > rail	truck < rail
truck < rail	truck > rail
truck < rail	truck < rail

thousand pounds per mile for all destinations beyond that to which the transport company travels. This method of computing and quoting freight rates economizes on the accounting effort of the transport firm.

TRANSPORT COSTS IN PRACTICE

Several economic geography texts contain data on the cost of transport. One example is Lowe and Moryadas (1975, pp. 31-39). But by far the most detailed published calculation is that of Harris and Hopkins (1972), which is based on 1965 data published by the Interstate Commerce Commission of the United States. These data show the out-of-pocket costs of transporting paper containers by rail and by truck for various regions of the United States. Here the rates for the Mid-Atlantic region are presented.

The data are contained in Figure 4.5. All the curves reflect out-of-pocket expenses only and so do not include markups for profits and overheads. The rail freight costs include terminal costs ($85.09 per car plus $0.039 per ton) and line haul costs. The line haul costs are broken at 32 miles, reflecting the hypothetical situation in which the load is picked up and carried for its first 32 miles by a local train ("way train") and then sent on by a "through train." Line haul costs are $0.289 per car plus $0.004 per ton for each mile by way train and $0.211 per car plus $0.002 per ton for each mile by through train. An allowance of $0.076 per ton is added for loss and damage. Rail freight rates, then, are essentially linear in distance. Truck rates are computed from common carrier costs. The overhead charges include pickup and delivery, platform handling, and billing and collecting; they are $3.68 per ton for greater than ten ton lots and $7.37 per ton for five ton lots. Line haul rates depend on both the distance travelled and the weight of the shipment; for 9 miles or less, the cost is $0.077 per ton per mile for five ton lots and $0.0616 per ton per mile for greater than ten ton lots. In contrast to rail freight charges, truck rates decline with distance. The net effect of these different costs is that trucks are cheaper than rail freight over short distances and particularly for small lots: For ten-ton lots,

54

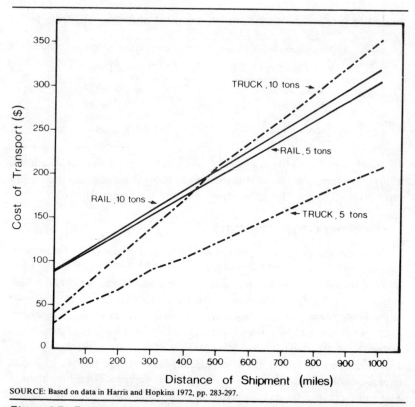

SOURCE: Based on data in Harris and Hopkins 1972, pp. 283-297.

**Figure 4.5 Freight Rates in Practice: Rail and Truck Costs for Paper
Containers in 1965**

rail becomes cheaper over truck for distances over 500 miles whereas for
five-ton lots, truck is cheaper than rail over all distances examined.

Least Transport Cost Location

This section tackles the first part of the problem of deducing some general
rules about location patterns. It does so by examining the nature of locations that minimize transport costs. Not only is the location problem
simplified by examining only a single factory, but the only factor examined is transport costs. Not only is the location problem simplified by examining only a single factory, but the only factor examined is transport
costs. By and large, too, the location of market and suppliers is assumed
to be given.

Figure 4.6 is a simplified version of Figure 4.2. Only one supplier of
each input is considered and transport rates are assumed to be simply pro-

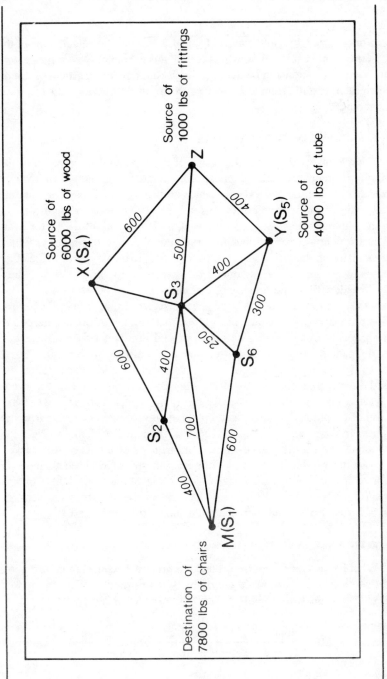

Source of 6000 lbs of wood

X (S₄)

Source of 1000 lbs of fittings

Z

Source of 4000 lbs of tube

Y (S₅)

600

500

400

400

S₃

300

600

400

250

S₆

600

S₂

700

400

Destination of 7800 lbs of chairs

M (S₁)

Figure 4.6 Simplified Location Problem

55

portional to weight and distance. Table 4.3 sets out the calculations of the least transport cost location, which is S_3 in this example. Now some general rules about location can be deduced by asking how the least cost location changes if certain technical changes occur in production and transport technology.

EFFICIENCY IN USING ONE MATERIAL

What happens if the weight of one of the inputs falls compared to the weight of other materials? Suppose, for example, that the old technology used 1,900 pounds of wood to produce 600 chairs. You may compute the least transport cost location for this old technology to check that it is S_4. In the old technology, when wood was inefficiently used, the factory was located at the source of that material; but the new technology, which economizes on the use of wood, allows the firm to locate nearer to the other inputs or the market. This conclusion is relevant to two practical issues.

Consider, first, the process of making steel slabs out of iron ore. This process is in practice divided into several stages. First, iron ore is mined. This ore may be 5 percent iron, say, so that 20 tons of ore are needed to produce 1 ton of iron. So the ore is treated in a beneficiation plant that makes pellets having an iron ore content of perhaps 70 percent. Thus 1 ton of pellets needs 14 tons of ore. In new technologies, these pellets are used directly to make steel: Ignoring other components, 1 ton of steel needs 1.4 tons of pellets. In the first process, 14 tons of inputs are needed to make 1 ton of output and so beneficiation plants are located at or near iron ore mines. In the second process, 1.4 tons of input are needed to make 1 ton of output and so steel plants are located near other inputs or the market.

Also, the production processes of many industries have been altered over time so as to economize on the use of a major input material. If the economy is sufficiently great, the pattern of location of the industries may change as factories are less attracted to those inputs than in the past.

EFFICIENCY IN MATERIALS PROCESSING

What happens if the entire production process becomes more efficient, so that more goods are made out of every ton of inputs? Weber proposed an index to measure this efficiency. His material index (MI) is

$$MI = \frac{\text{total weight of localized materials used up}}{\text{weight of finished product}}$$

TABLE 4.3 Calculations of Transport Costs for Problems of Figure 4.6

Site of Production	Transport Costs to Site from Source			Transport Costs to M	Total Cost of Transport
	X	Y	Z		
S_1	6000	3600	1200	0	10800
S_2	3600	3200	900	3120	10820
S_3	2100	1600	500	5460	9660
S_4	0	3000	600	7800	10860
S_5	4500	0	400	7020	11920
S_6	3600	1200	700	4680	10180

where *localized* raw materials are those not available at every site. In our example, power and water are not localized (they are *ubiquitous*). The data of Table 4.3 reveal that the material index for this problem is

$$MI = \frac{6000 + 4000 + 1000}{7800} = 1.41$$

Given this definition, we can rephrase the question.

What happens if the material index falls? Suppose, for example, that a newer technology allows 750 chairs of 13 pounds to be produced from the given inputs. The material index has fallen to 1.14. The least transport cost location is now S_1. We conclude, then, that as the material index falls, so production is attracted toward markets.

Now, two conclusions can be drawn. First, at any time, those industries in which the material index is lowest should be located nearest their markets whereas industries with a high material index should be located nearer their dominant raw material. This prediction can be compared to reality; in extreme cases, it is true. Soft-drink factories use about an ounce of cordial and a 4 ounce can in conjunction with an ubiquitous material (water) to produce a 12 ounce can and 4 ounce container of soft drink; its material index equals $1 + 4 / 12 + 4 = 0.31$, which is less than one. Therefore soft-drink canning is a market-oriented industry. Primary metal smelting is a converse example. In intermediate cases, the location of industries conforms less well with our prediction because other factors (to be considered in the next chapter) are important. Second, over time as materials-processing technology has become more efficient, the average material index in manufacturing has fallen, and so industry should have become less and less dependent on raw material sites and more attracted to markets. This prediction is remarkably well borne out in aggregate. In the nineteenth century in Britain and in the United States, coal fields and iron ore fields became the locations of such major industries as iron and steel. However, the newly discovered ore deposits of Australia and Minnesota and the coal fields of British Columbia and Australia have not become the sites of new iron and steel plants. Similarly, the new areas of aluminum or uranium do not attract production in the way that coal and iron once did.

COMMODITY SPECIFIC TRANSPORT RATES

What is the effect of charging different transport rates for different commodities? Suppose for example, that because chairs are bulky in relation to their weight and must be packed carefully, they are charged $1.25 per 1,000 pounds per mile (compared to $1.00 for inputs). Table 4.4 presents the new calculations. You should compare these calculations with the ones you made in the previous section for the production of 750 ($= 1.25 \times 600$) chairs when all transport rates were $1.00 per pound per mile. These two calculations are the same. That is, products that are expensive to transport per ton because they must be packaged or are perishable act as though they were heavier than they really are.

There are two good examples of this phenomenon. Vegetables are highly perishable, especially if they are to be frozen in peak condition. Hence the inputs are expensive to transport to freezing plants. Therefore those plants are located as if the weight of vegetables were very high and the material index high too—near the source of inputs. By contrast, bread uses flour and a few smaller ingredients and incorporates a little water. The bread, however, is more perishable than the inputs and is therefore expensive to transport. Consequently, the baking industry is located as if the bread were heavy in relation to its inputs, namely its markets.

CHANGING TRANSPORT RATES

What happens if all transport rates fall? Suppose that you redo the calculations of Table 4.3 with all transport rates lowered ($0.50 per 1,000 pounds per mile). You will see that the least cost location is unchanged. This is important: The reduction in transportation costs that has occurred over the last two hundred years has not affected the least cost transport cost locations. (We shall observe other effects of this change later.)

Your calculations will reveal one other effect of halving the general transport rate. Even though the least transport cost location is not altered in any way, nevertheless the absolute magnitude of the advantage of S_3 over other locations has been halved: Whereas S_1 used to cost $1,140 more than

TABLE 4.4 Calculations of Transport Costs when Chairs Are Charged More

Site of Production	Transport Costs to Site from Source			Transport Costs to M	Total Cost of Transport
	X	Y	Z		
S_1	6000	3600	1200	0	10800
S_2	3600	3200	900	3900	11600
S_3	2100	1600	500	6825	11085
S_4	0	3000	600	9750	13350
S_5	4500	0	400	8775	13675
S_6	3600	1200	700	5850	11350

NOTE: Chairs cost $1.25 per 1000 pounds per mile.

S_3, it now costs only $570 more (and is similarly reduced for other locations). Thus, reducing the transport rate does not alter the least transport cost location but it does reduce the magnitude of variations in transport costs between different locations.

TERMINAL COSTS

What is the effect on location of transport terminal costs? Suppose that in addition to line haul costs a terminal charge is added to transport costs, equal to $500 plus $0.10 per pound. Table 4.5 shows the effect. The three intermediate locations (S_2, S_3, and S_6) become more costly compared to input or market sites: If the factory is located at an intermediate site, four terminal costs must be paid, whereas if the factory is at an input or market site, only three terminal costs have to be paid. That is, terminal charges force factories to locate at material sources or at the markets rather than at intermediate sites.

What, then, has been the locational impact of the truck, which has low terminal costs compared to railways? In the nineteenth century, the rail system had high terminal costs, which forced firms to locate at input sites or at markets. The truck has reduced terminal costs, thus freeing firms from the compulsion to locate at input or market sites. The theory predicts that the innovation has allowed firms to choose intermediate locations rather than input or market sites. In one respect this prediction is borne out in reality: The truck has allowed manufacturing industry to leave the major metropolitan areas and to locate in small towns and rural areas.

CHOICE OF SOURCES

These general conclusions apply to the location decision for given sources and markets. They apply, however, to any *any* set of sources and markets,

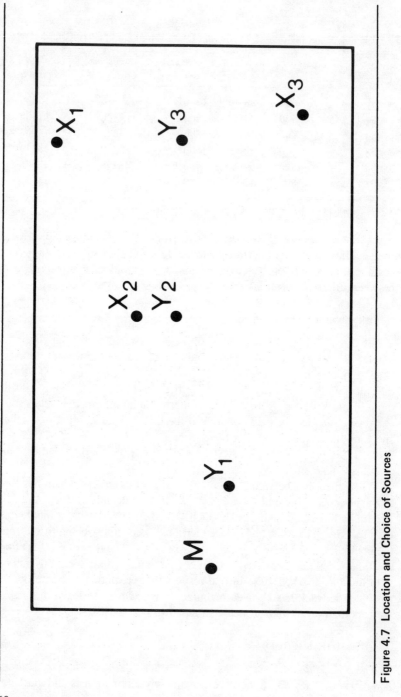

Figure 4.7 Location and Choice of Sources

TABLE 4.5 Calculation of Transport Costs with Terminal Charges

Site of Production	Transport Costs to Site from Source			Transport Costs to M	Total Cost of Transport
	X	Y	Z		
S_1	7100	4500	1800	0	13400
S_2	4700	4100	1500	3400	14700
S_3	3200	2500	1100	6740	13540
S_4	0	3900	1200	9080	13640
S_5	5600	0	1000	8300	14900
S_6	4700	2100	1300	5960	14060

NOTE: Terminal charges are $500 + $0.10 per pound.

and thus they apply to all such sets. Hence, these conclusions are quite general and independent of the firm's choice of sources and markets. Consider, then, what input sources should be used to produce for a given market. In some special cases, the principle whereby this problem is solved can be augmented by general rules.

Suppose that a factory's material index is less than 1.0. Where will it locate and what will be its input source? The lower the material index, the more a factory is attracted to its market. And if MI < 1.0, then the least transport cost location must be M. You can check this with your own examples. So, whatever input sources the firm employs, its factory must be located at M. It follows, of course, that the least transport cost sources are those that are nearest M, that is, nearest to the factory.

Suppose, second, that a factory's material index exceeds 1.0 and that there is one dominant input: The factory must locate at or near a source of that material. Then its source is chosen to be the one nearest the market; but the sources of other materials are chosen to be near the location of the dominant one. Consider, for example, Figure 4.7. There are three sources of the input X, near one of which the factory must locate. If the factory has to be near X_1 or X_2 or X_3, the transport costs to market are minimized if the factory locates near X_2. But of the three input sources of $Y—Y_1$, Y_2 and Y_3—the factory uses not Y_1 (the source nearest the market) but Y_2 (the source nearest X_2). Caution should be exercised, however, as this is not a hard and fast rule.

Transport, Plant Size, and Location

The previous section has assumed that the factory size is given (the market, too); but any investment decision must incorporate an analysis of the maximum profit size of a factory. This chapter concludes with such an analysis. First, the notion of economies of scale is introduced. Second, the choice of factory size is analyzed, initially ignoring transport costs and then adding transport costs to the analysis. (A more formal approach to the analysis of plant size and location is contained in King, 1984).

ECONOMIES OF SCALE

The phrase *economies of scale* means the savings in production costs as operations are enlarged. There are two kinds of economies of scale. The first are the economies of large firms over small, some aspects of which were examined in Chapter 2. The second kind of scale economy is that of large factories; it is this economy that is analyzed in this section. In theory there are several reasons why large factories are more efficient (produce at less cost) than small ones.

The first reason is the division of labor within a factory. A small factory that employs only one person to perform all the operations needed to produce a commodity is using unspecialized labor. The person spends, say, 30 minutes on task 1, 2½ hours on task 2, 2 hours on task 3, 3 hours on task 4. Suppose, however, that the factory hires 97 people: 60 for task 1, 12 for task 2, 15 for task 3, and 10 for task 4. Then each worker could learn one operation thoroughly, no labor time would be lost in passing from one operation to the next, each operation could be mechanized more easily, and each more specialized laborer would be less skilled in aggregate and therefore paid less. In this example, 97 is the least number of employees in a factory where each laborer was specialized in this way. The division of labor was one of the bases for factory organization in the nineteenth century, before mechanization.

Now that factories are generally mechanized, a second economy has become important: the economy of large machines. Equipment that stores materials and machines (the power of which depends on their volume) are subject to a geometric economy: The volume of a cube increases as the cube of its linear dimension but the area of its faces increases as six times the square of that length. Thus, ships, storage tanks, and factories are all subject to this economy, the only limitation on which is provided by the strength of materials.

The third main economy is the economy of massed reserves. Factories maintain inventories of spare parts for machines and reserve stocks of materials to guard against breakdown or emergencies. Now the size of these reserves must increase as the level of production increases; but for a given risk, the level of reserves rises more slowly than the level of output. Hence the cost of the reserve per unit of output is less for large levels than for small.

These three economies are of differing significance. The economy of reserves is real but probably quite small. The economy of large machines is most important for factories in which storage is important but is subject to the limits of material strength. The main economy of large scale is that which arises from the division of labor and the need to link processes and machines in a continuous process. However, there exist no convincing measurements that demonstrate the real existence of scale economies, at least beyond a minimum size. (See Lloyd and Dicken 1977, pp. 273-280, for further discussion of this issue.)

FACTORY SIZE

Suppose that a factory is going to be built to produce a commodity that will be sold in a city. Several factories already exist in the city but the entrepreneur believes that there is room for another. Initially, assume that the costs of transport do not vary with the level of output. How large should the factory be?

Figure 4.8 illustrates the method of analyzing this question. The solid line on this diagram shows how costs vary in relation to scale: The hypothesis is that there are some economies of scale, but for output levels greater than 1,000 units per month scale economies are insignificant. (You can perform the same analysis with different hypotheses about scale economies.)

The dashed line of Figure 4.8 reflects the entrepreneur's beliefs about how much can be sold at each given price. If the price is too high, few people will buy; as the price is reduced, more will buy, until at a very low price, the factory can capture the entire market in the city.

The firm's strategy is to try to maximize its profit per unit of output—that is, to choose a level of output such that the difference between the price received and the cost paid is greatest. On Figure 4.8 this occurs at an output of 750 units per month, which is below the minimum efficient size of the plant.

Now let us introduce costs of transport. If the factory is to sell to one city, transport costs are small. As it increases the scale of its operations, however, it must sell to people who live further and further away. Hence the larger the output of the firm, the greater the costs of transport it must pay to deliver its output to the consumers. The firm can now increase its output by selling more to its local market (by lowering its price) or by increasing the spatial extent of the market it serves.

Figure 4.9 shows how to analyze this problem. The solid lines show the costs of production and transport to market for market areas of varying

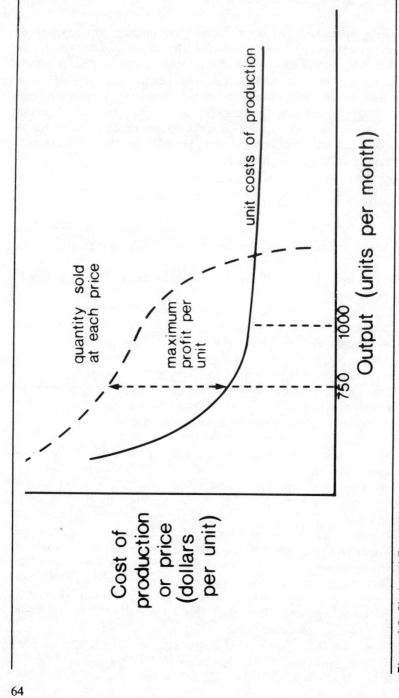

Figure 4.8 Choice of Factory Size

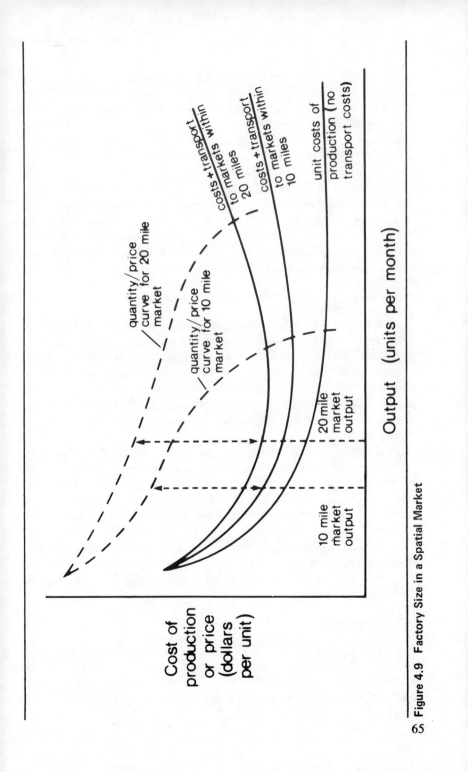

Figure 4.9 Factory Size in a Spatial Market

65

extent: The larger the market, the greater these costs. The dashed line shows how much the firm can sell for a given price in each of the market areas. Hence for each market area, an optimum level of output can be determined to maximize the price-cost difference. In this case you can see that the firm makes more profit if it sells to a market with a radius of 20 miles than if it sells over a 10-mile radius.

Now the figure also allows you to analyze the effects of changes in transport rates on the size of factories. Suppose that transport rates fall. The quantity/price relationship is not affected, but the costs of serving larger markets fall. Hence the effects of transport costs on the cost curves fall, and the higher curves diminish toward the costs (no transport) line. It follows that the benefit of selling to larger markets is increased, and so plant size rises as transport costs fall.

Conclusion

This chapter has analyzed the structure of transport rates and examined the relations between transport costs and location. By and large, transport rates have fallen over the last two hundred years, and the least cost theory predicts that this change has affected location patterns in several ways. These effects include the increasing size of markets served by one factory, the greater freedom of factories to locate at intermediate sites (neither at material sources nor at markets), and the decreasing attraction of industry to raw materials.

In an immediate sense, then, changes in the structure of transport costs cause these locational adjustments. Yet you should beware of a mechanical or technological interpretation of locational change as due to changing transport cost structures and materials processing technology: The new technologies have the observed locational effects only in the context of the specific social system (e.g., in the context of profit maximizing firms); and the new technologies do not appear out of nowhere—they have to be produced as a result of investment decisions by entrepreneurs—and thus the causes of the technical changes and their specific form need to be investigated.

Further Reading

One of the best discussions of transport costs and location that could be used to follow up this chapter is contained in *Location in Space* by P. E. Lloyd and P. Dicken, (1977, pp. 157-195). P. Haggett, A. D. Cliff, and A. Frey discuss transport networks in *Locational Models* (1977, pp. 64-96).

5. LEAST COST THEORY: PRODUCTION COSTS AND AGGLOMERATION

In addition to transport costs, least cost theory examines the influence on location of two other factors: spatial variations in production costs and economies of agglomeration. This chapter completes the presentation of the theory by examining these two factors. The discussion is divided into five parts. The following section presents the basic theory about the effects of production cost variations on location. This theoretical discussion is succeeded by three more empirical sections that examine nonwage costs, labor costs, and the issue of amenities and business climate. This chapter concludes by reverting to a theoretical mode and examining the effects of agglomeration on location.

The last chapter ended by warning against technological determinism, for transport rate changes have to be produced by society. This idea, that spatial variations in production costs are themselves produced by society, is more emphasized in this chapter. The central claim of the chapter is that labor costs are becoming the most significant factor affecting location but that the spatial variations in these costs are themselves the outcome of earlier development decisions. In this sense, labor is a produced factor of production.

Production Costs

Not only do transport costs vary over space, but so do production costs. For any given market and suppliers, the choice of location must therefore reflect spatial variations in production costs as well as in transport costs. In practice, this amounts to (1) discovering the production costs at each possible location; (2) adding production and transport costs at each location; and (3) choosing that location at which these combined costs are least. This is quite straightforward—but can we discover some general principles about the effects of production costs on location?

One important principle can be deduced if we return to Figure 4.7 and Table 4.3. Suppose we add to that problem the data on production costs for 600 chairs at each site (see Table 5.1). The least cost location is site S_3, for its combined transport and production costs are less than those for all other sites. Now suppose that the transport cost is halved. Then the site-to-site variation in transport costs is halved too, and spatial variations in production costs take on greater significance. As Table 5.1 shows, the new least cost location is S_2. (Observe that S_2 is not, however, the site with the least production costs.) The principle is this: As transport rates fall in relation to costs of production, so production cost variations rise in significance as a locational factor compared to transport costs. We can apply this principle in two ways.

First, we know that transport rates have fallen over time in relation to other costs. We shall also see in the following sections that there remain large spatial variations in costs of production—especially wages. Hence we anticipate that this relative fall in transport rates has permitted firms a greater freedom than they used to have to seek out least production cost locations. In particular, this means that firms can now more readily escape the regions in which labor costs are high.

Second, costs of transport vary for different industries. Some industries sell a light product of high value, for which transport costs are small compared to production costs: fashion clothing for example. Other industries use bulky materials and produce bulky output and so are dominated by transport costs: iron and steel, for example. Our principle tells us that different industries locate with respect to different factors: Production costs dominate for fashion clothing; transport costs dominate for iron and steel.

Nonlabor Costs

There are many other costs of doing business besides transport costs and labor costs: Plant and equipment have to be acquired; land has to be be bought or rented; and taxes must be paid. Other costs, like the cost of the sales effort, are presumed not to vary with the location of the factory. This section examines the spatial variability in the cost and availability of these factors.

PLANT AND EQUIPMENT

Some factories must be located in the middle of nowhere—for example plants that pelletize iron ore, or process lumber. Such factories are built far from existing road or rail links. Hence the plant, equipment, and the transport link itself are all expensive to build. Most factories, however, are not built in such isolated places—because there are no workers and plants would be expensive. Being input-output operations, factories generally must be near existing transport links. For such factories, spatial variations in the costs of plant and equipment are insignificant.

Consider the following data (for 1978, from U.S. Department of Commerce *Statistical Abstract of the United States* 1981, p. 777). In that year, U.S. manufacturing plants spent $55.2 billion on new capital expenditures. Suppose that these capital items were in such out-of-the-way spots that an extra 10 percent had to be spent on transport: $5.52 billion. Now, the yearly

TABLE 5.1 Effects of Production Costs on Location

Production Site	Total Cost of Transport from Table 4.3	Production Costs	Transport Costs After Rate Is Halved	Transport plus Production Before Rate Is Changed	After Rate Is Halved
S_1	10800	2150	5400	12950	7550
S_2	10820	2000	5410	12820	7410
S_3	9660	2600	4830	12220	7430
S_4	10860	2000	5430	12860	7430
S_5	11920	1640	5960	13570	7600
S_6	10180	2510	5090	12690	7600

interest at 12 percent on these transport costs is $0.66 billion, which is merely 0.1 percent of the value added by manufacture and only 0.54 percent of the total profit earned by manufacturers in 1978. Thus spatial variations in the price of plant and equipment can have little effect on aggregate locational choices within advanced economies (although they may have greater effect within particular industries).

However, the effective cost of plant and equipment could vary spatially if there are spatial differences in the cost of the capital that has to be borrowed to buy the plant and the machinery. This cost varies spatially only if there are barriers to mobilitiy—that is, if a lender in New York, say, is more willing to lend to a New York firm than to an Atlanta firm. There are two such barriers. The first is a difference in information. When travel was costly and slow, economic conditions and investment potential in borrowing regions were not widely appreciated in lending regions—hence, lenders were less willing than they ought to have been to lend outside their home region. The second barrier is the cost of the loan made between separate places. In the nineteenth century, New York banks made loans to Westerners via intermediaries, who charged a fee: So Western loans cost more to administer than loans to North Easterners. Nowadays, the major cost of loans is the risk of lending capital across national boundaries. Within the United States, a creditor is subject to law and nonrepayment brings legal penalties. But American firms that invest abroad have no such legal backup, for foreign firms cannot be required by U.S. law to honor a debt.

Hence, the cost and availability of capital do vary spatially (Lloyd and Dicken 1977, pp. 230-231, present some evidence, too). Of course, anyone would find it hard to borrow in order to build a steel mill in southern Florida: This however, is hardly an effective restriction. The effective locational restrictions that arise from capital are caused by lack of information and the costs of distance. Within advanced economies, these restrictions are now of little importance, but they do deter cross-boundary flows.

LAND AND TAXES

There are enormous variations in the price of land between the center of a metropolitan area and its periphery. Land at the center can be thousands of times as costly as peripheral land. These variations are a strong incentive for firms to locate in suburbs rather than in metropolitan areas if they are able to. But the variation in prices between land in the suburbs and land in small towns is much less.

How significant are these variations? In the United States in 1978, the average factory had a net value of $10,993 per worker in structures, of which, say, one-third is land—about $3665 per worker. If the firm paid interest on this investment at 12 percent per annum, the cost of land is $440 per worker, less than 1 percent of the value added per worker and only 0.4 percent of the value of shipments per worker. Thus while the difference between central city and suburban land prices is important (and this is one of the reasons why manufacturing plants have been decentralizing to the suburbs since World War II), the difference between the prices of land in different cities or between different areas is not great enough to affect profitability much.

The same is true of taxes and business incentives. We frequently learn that tax levels are too high or business incentives too low for a firm to locate a factory in this place or that—so we are led to believe that such government behavior can affect location. Yet we should be leary of such claims; firms have an incentive to claim that taxes or incentives are more influential than they really are in order to obtain a better deal for themselves. So we must consider the level of taxes and incentives more carefully.

Governments in most Western European and North American countries offer incentives if firms locate their factories in particular places, such as depressed areas. Such incentives typically take several forms: reduced business tax obligations; low cost business loans; overhead projects (such as business parks); and freedom from legal restrictions on such business practices as labor relations. Businesspeople strongly favor such incentives, and popular impressions suggest that location incentives are important influences on location choice. Yet the evidence is uniformly negative: Firms make their plans, find out about the incentives available at the planned location, and then apply for the incentives (Harrison and Kanter 1978); econometric models find no statistically significant relationship between growth in manufacturing employment in the U.S. states and the level of state and local taxes (Bloom 1955; Thompson and Mattila 1959; Treysz et al. 1976). Thus you should remain skeptical about the role of incentives. While businesses obviously benefit from them, the incentives cannot be large enough nor continue long enough to affect firms' locations in the long run.

CONCLUSIONS

The most important point emphasized in this section is that nonlabor costs do not vary greatly over space—certainly not sufficiently to affect profitability noticeably. The prices of plant, equipment, land, and taxes all vary spatially. However, the cost of the assets to the firm is not their price but the costs of the interest needed to pay for the loan to buy them, and generally this cost is only a small proportion of sales or even of value added. Apart from the costs of transport, then, the only remaining factor of location is labor costs.

Wage Rates

Labor is the essential factor in manufacturing. The cost of labor is high and the evidence from political advocates of wage controls and the seriousness of wage bargaining suggests that business people regard this cost as significant. Equally, it has been argued that unionization and high wage rates in the northeastern United States have contributed to the industrial decline of that region. So now we examine data on the spatial variability of the costs of labor.

WAGE RATE VARIATIONS

Despite mechanization, automation, and computer processing, labor costs remain high. In 1977, payrolls accounted for 45.1 percent of the U.S. value added (U.S. Department of Commerce, *Statistical Abstract*, 1981, p. 777). In some industries, the payroll is a small proportion of value added: only one-sixth in tobacco products (SIC 21) and petroleum and coal products (SIC 29), abut 25 percent in chemicals (SIC 28), and 33 percent in food industries (SIC 20). In other major SIC groups, the payroll exceeds 40 percent of value added, reaching 50 percent in furniture (SIC 25), leather (SIC 31), and primary metal industries (SIC 33). In U.S. manufacturing as a whole, a 20 percent increase in payroll would absorb 9 percent of the value added. Unlike the costs examined in the previous section, the costs of labor are a high proportion of the costs of manufacturing.

The costs of the payroll understate the labor costs that firms must pay. In addition to wages and salaries, U.S. firms make contributions to employees' and other social security programs. In 1977, these contributions added $50 billion to the payroll bill, equivalent to 8.5 percent of total value added (U.S. Department of Commerce, *Census of Manufactures 1977*, General Summary, p. 1-46). Thus even now, labor accounts for over half of U.S. value added in manufacturing.

The payroll itself can be divided into two parts. One is the payment to production workers (wages) and the other is the payment to administrative, sales, research, and other nonproduciton workers. Production workers earned in 1977 about 60 percent of the payroll, a fall from 68 percent in 1954 and 75 percent in 1947 (U.S. Department of Commerce, *Statistical Abstract* 1981, p. 777).

Not only are labor costs high, they also vary over space. Figure 5.1 is a map of the payroll per employee in each state of the United States as a proportion of the U.S. average in 1972. The wage rates in the areas with highest earnings (Delaware, Michigan, and Washington) are all more than 50 percent greater than in the states with lowest wages (North Carolina, Mississippi, Arkansas, New Mexico). In general, wages are below the U.S. average in New England and throughout the South except for in Connecticut, Delaware, Maryland, and Washington, D.C.; wages are above the U.S. average in the Mid-Atlantic, East North Central, and Pacific regions, except for Pennsylvania. These large spatial variations conform to the popular perception.

But these variations in wages are not the ones that most concern a firm. The wage rates shown on Figure 5.1 may vary for two reasons: Either Michigan's firms pay higher wages in each industry than do Arkansas's firms, or Michigan's employees work in higher wage industries than do Arkansas's employees. Tables 5.2 and 5.3 provide some evidence about these alternatives for two industry groups—textiles (SIC 22) and paper industries (SIC 26). Even within the three-digit industries, there remain substantial interregional variations in hourly wages. Within the textile industries, the Pacific, Mid-Atlantic, and East North Central states pay above the U.S. average, and the South pays below. In the paper industries, the pattern of wage variations is less consistent: The North East generally pays below and the West pays above the U.S. average, but the North Central and Southern states sometimes pay more and sometimes less than the U.S. average. Even within one industry, then, there are large variations between states in the average wage paid per worker.

CAUSES OF WAGE RATE VARIATIONS

A complete theory of industrial location cannot simply accept the evidence of Figure 5.1 and Tables 5.2 and 5.2, but must inquire about the origin of spatial differences in wages. Wage levels are not simply given phenomena but are produced by social processes. How do spatial variations in wage rates arise? Why do they persist? We can only touch on these issues here.

A popular reason given for variations in wage rates over space is the degree of unionization. It is thought that the rate of union membership is highest in the Northeastern region of the United States—the old manufacturing belt—which is one of the high wage areas. By contrast, the low wage South has low rates of unionization. Yet doesn't this seem too simple? In 1982 when the auto manufacturers were in financial trouble, they persuaded their highly unionized workforce to accept wage and benefits cuts, just as Chrysler had done earlier when it was faced with bankruptcy. Since its fiscal crisis, New York City has imposed drastic real wage cuts on its employees.

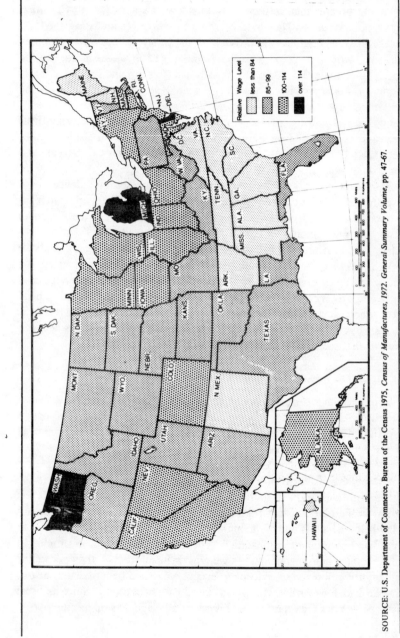

SOURCE: U.S. Department of Commerce, Bureau of the Census 1975, *Census of Manufactures, 1972. General Summary Volume*, pp. 47-67.

Figure 5.1 Average Wage per Hour of Manufacturing Employees

TABLE 5.2 Wage Rates (dollars per hour) by Census Division for Textile Industries, 1972

Census Division	SIC 22	SIC 221	SIC 222	SIC 223	SIC 224	SIC 225	SIC 226	SIC 227	SIC 228
United States	2.79	2.78	2.82	2.85	2.57	2.70	3.00	2.95	2.59
North East	3.06	2.73	3.05	2.92	2.65	3.02	3.48	3.46	2.61
New England	2.94	–	3.06	2.78	2.63	2.86	3.23	–	2.58
Mid-Atlantic	3.12	–	3.03	3.39	2.67	3.05	3.66	–	2.66
North Central	3.32	–	2.73	3.00	2.67	2.97	2.91	3.28	2.47
East North Central	3.41	–	–	–	2.82	3.01	–	–	–
West North Central	2.81	–	–	–	2.25	2.84	–	–	–
South	2.68	2.78	2.79	2.71	2.44	2.52	2.73	2.82	2.59
South Atlantic	2.70	2.80	2.80	2.70	2.44	2.54	2.75	2.85	2.60
East South Central	2.60	2.68	–	–	–	2.46	2.55	2.70	2.55
West South Central	2.54	2.71	–	–	–	2.44	2.50	2.53	2.38
West	3.02	–	3.13	3.00	2.75	2.63	3.00	3.58	2.56
Mountain	2.42	–	–	–	–	2.38	–	–	–
Pacific	3.08	–	–	–	2.75	2.70	–	–	–

SOURCE: U.S. Department of Commerce, Bureau of the Census. 1975. *Census of Manufactures, 1972.* vol. 2, part 1, pp. 22-3 to 22-7.

NOTES: Missing data are not available for nondisclosure reasons.
- SIC 22: textile mill products
- SIC 221: weaving mills, cotton
- SIC 222: weaving mills, synthetics
- SIC 223: weaving and finishing mills, wool
- SIC 224: narrow fabric mills
- SIC 225: knitting mills
- SIC 226: textile finishing, except wool
- SIC 227: floor covering mills
- SIC 228: yarn and thread mills

Perhaps unions are as powerful as the financial strength of employers allows them to be.

The second theory of wage rates is a cost theory. This claims that the local wage rate must in the long run be set at the level that will reproduce the labor force there—that is, will bring on new laborers of the appropriate skills and training to replace those who retire or leave, and will sustain each existing worker at the accepted level of living. On these grounds, those with longer training should be paid more and those who live in more expensive places (like the North and big cities) should be paid more. But this theory leaves open what is an "accepted level of living."

Third, there are elements of supply and demand. Areas of declining industries—or industries that are shedding labor—provide labor that is seek-

TABLE 5.3 Wage Rates (dollars per hour) by Census Division for Paper Industries, 1972

Census Division	SIC 26	SIC 262	SIC 263	SIC 264	SIC 265
United States	4.15	4.83	4.86	3.79	3.72
North East	3.90	4.45	4.31	3.67	3.62
New England	3.95	4.44	4.41	3.66	3.39
Mid-Atlantic	3.87	4.45	4.26	3.67	3.69
North Central	4.17	4.87	4.58	3.96	3.86
East North Central	4.17	4.89	4.56	3.89	3.86
West North Central	4.17	4.78	5.11	4.20	3.84
South	4.11	4.95	4.94	3.49	3.38
South Atlantic	4.07	4.68	4.90	3.53	3.48
East South Central	4.07	5.04	4.93	3.40	3.28
West South Central	4.21	5.21	5.06	3.54	3.36
West	4.97	5.89	5.46	4.36	4.51
Mountain	4.23	–	–	3.43	–
Pacific	5.03	–	–	4.47	–

SOURCE: U.S. Department of Commerce, Bureau of the Census. 1975. *Census of Manufactures, 1972.* vol. 2, part 1, pp. 26-3 to 26-6.

NOTES: Missing data are not available for nondisclosure reasons.
SIC 26: paper and allied products
SIC 262: paper mills excluding building paper
SIC 263: paperboard mills
SIC 264: miscellaneous converted paper products
SIC 265: paperboard containers and boxes

ing work. In the recent history of the United States the major industry of this type has been agriculture, and the major farming area that has lacked rapid industrialization has been the South. Hence Southern farms have been producing labor, which has been seeking employment. Some Southerners migrated North, but those who stayed have provided the basis for Southern low wages. By contrast, the expanding industries of the Pacific and North Central regions have had to attract labor from other regions—that is, offer higher rates of pay. The rate of migration of workers has not been sufficiently high to offset wage differentials because the benefits from migration (a job, a higher wage) are at least partially offset by the costs of migration (cost of moving, personal disruption, low house prices in depressed regions). So declining industrial regions continue to offer cheap labor.

This argument envisages a dynamic spatial relationship between wage levels and location. Growing areas need to attract labor and so must pay high wages; depressed regions have a surplus of labor and so offer low wages. Once the disparity in wage costs between high wage and low wage areas becomes sufficient, industrial firms have an incentive to try to locate in the low wage areas. In other words, we must be careful not to regard wages as a purely ''independent variable.'' It is not merely that wages affect location but also that location choices (i.e., regional economic development) affect wages.

Amenities and Business Climate

The previous sections have tried to measure some of the production costs that vary over space. Yet some attributes of locations are less easily measured quantitatively. For example, it has sometimes been claimed that some private firms are located in places where their owners prefer to live, and that the growth of the South and Southwest is partly attributable to its climate. Can this argument be assesed? Similarly, business people do assess places in terms of their receptivity to the ideals of business, and some people claim that some places have a more favorable *business climate* than do others. What is meant by business climate, and is it important? This section addresses these two questions.

AMENITIES

In some respects the South is a cheaper place to live and to do business in than is the Northeast. Heating costs less, property taxes are often lower, and wages and statutory labor practices cost less in the South than in the Northeast. But these factors have already been considered in the list of location variables. The question of amenities, then, is this: Are there any pure (nonmonetary) effects of climate, landscape, intellectual life, or other amenities that attract firms? Put in this way, the question has only one answer: Such attractions can only affect a firm's choice between two otherwise equally profitable locations. Thus the pure effect of amenities must be small.

In practice, then, it seems that amenities can influence location decisions only through their effect on other costs. The presence of intellectual resources may make it easy for a laboratory to attract scientists. To the extent that sun and warmth reduce heating costs they diminish production costs. If workers do really prefer sun to snow, then it may be easy to attract labor to warmer spots. But these hardly seem to be major considerations in locating multimillion dollar factories.

BUSINESS CLIMATE

The concept of business climate refers to the willingness of the population of a place to subscribe to the ideals of business. In practice, this means giving the firm freedom to run its business with minimum interference from the state or from workers' organizations.

The state can interfere with business in several ways. It can raise property taxes or, conversely, can offer locational incentives. The state can also regulate industry: by imposing legal limits to pollution, by imposing employment standards, and by requiring certain forms of health and retirement insurance. Workers' unions may affect wage levels, employment standards, and the freedom of the firm to lay off, transfer, or discharge workers.

In the first instance, at least, business climate probably varies inversely with wages—being most favorable to business where wages are relatively low, or workers least unionized. However, business climate is not an independent locational variable; like wages, it is produced by the existing industrial structure. That is to say, where the demand for labor is high in relation to its supply, workers have more power than where there are few jobs available. Hence, labor in high demand regions can more easily force union recognition, employment and health standards, and pollution controls. Thus we can hardly claim that business climate affects location without explaining how business climate arises in response to existing levels of manufacturing development.

Agglomeration

An *agglomeration* is a point on the earth's surface at which economic activity is particularly dense. A metropolitan area is a spatial concentration of industry. *Economies of agglomeration* are those savings in production costs that occur when factories locate near one another. Suppose for example, that there are two possible sites, S_1 and S_2, and two factories, F_1 and F_2. Table 5.4 gives the production costs per unit of output for each factory depending on its location. Considered alone, F_1 should locate at S_1 and F_2 at S_2. But if F_1 and F_2 both locate at S_2, they each save some production costs: This saving is the economy of agglomeration. What are the sources of agglomeration economies?

Clearly, if the factories sell goods to each other, they can reduce transport costs by being adjacent. If a steel mill sells hot rolled steel to a nearby can-maker, both save on transporting the steel as well as on the cost of reheating it. If parts manufacturers sell components to a nearby auto factory, there will be savings on transport costs and on the personal travel needed to en-

TABLE 5.4 An Example of Agglomeration Economies

Production Costs (dollar per unit)		Site S1	Site S2
Factory F1 alone		10.47	12.63
Factory F2 alone		16.92	15.87
Factories F1 and F2 together	F1		10.36
	F2		15.21

sure adequate quality control. Law firms, accountants, and other business services cluster around firms' headquarters, saving on the transport costs and time needed for personal consultation.

All these advantages from proximity are savings on transport costs. They are not, therefore, agglomeration economies strictly defined; agglomeration economies are savings in production costs. Three such economies can be defined, although they are hard to measure.

The first potential source of agglomeration economies is the saving in costs due to scale. Some firms can specialize to provide a business service to a variety of customers—international bankers cannot efficiently serve a single business. There are also economies of massed reserves: As the total average demand for a service or part rises, the total level of inventory needed in case of sudden demands rises too, but at a slower rate. Thus a smaller proportion of capital is held as inventories by firms in large agglomerations than by dispersed firms. Particularly significant is the fact that because of the level of demand, transport services to large agglomerations are better developed, more frequent, and more competitive than is transport to small towns.

Second, there are economies of information associated with agglomerations. Business people can communicate with governments, research organizations, suppliers, customers, and competitors if they are all located nearby. These economies seem to apply mostly to two kinds of establishments. On the one hand, many headquarter functions of large corporations seek locations from which to make frequent personal contacts with governments, research organizations and business services. This economy is reflected in the concentration of headquarters functions in larger cities (see Table 2.4). On the other hand, small factory-firms in industries where demand is uncertain (because it is individualized, subject to fashion, or still changing technically) agglomerate to share information about demand: Examples of such industries are commercial printing, women's clothing, and the radio industry in the early part of this century. These economies are like transport costs, but they depend on the ability of people to meet rather than on the ability of goods to flow.

Perhaps, however, the main economy of agglomeration arises from a consideration of social fixed capital—the infrastructure of society. Items of social fixed capital are commodities, the consumption of which extends over several production periods; this includes railroads, highways, banks, commercial facilities, hospitals, schools, and shopping centers. These goods are long-lived, difficult to alter, and immobile. Some of these items are privately owned, but others (such as highways) are publicly owned. All, however, offer shared use to customers. Similarly, a less tangible element of fixed capital is the structure of society in a place, especially the relations between workers and managers. Thus within the United States there are quite distinct regional differences in degrees of unionization, control of industry, and other reflections of the worker-manager relation. This relation is also an element of the social fixed capital of a place: It is shared by all employees, it lasts a long time, and it requires investment (e.g., in schools) to maintain or to alter it.

The existing system of industrial production works by combining privately produced items (raw materials, factories, machinery, and so forth) with publicly produced items (which are primarily the social fixed capital). Yet the social fixed capital is evidently expensive to produce, and its costs are therefore minimized if agglomerations are produced in as few places as is consistent with spatial variations in transport costs and private production costs.

This view of agglomeration establishes the process as part of the system of industrial production. As an agglomeration is regarded as a conglomeration of social fixed capital (and its users), it is natural to envisage that capital becoming outdated. Just as private fixed capital—a machine or factory—can become obsolete, so can an agglomeration in the sense that other arrangements of social fixed capital can deliver their services to users more cheaply. The City of New York is an example. It offered the following as social fixed capital: free university system; unparalleled port facilities; a variety of business services; many small rentable factories; few freeways; a large skilled but highly unionized and well-paid labor force. These things are not needed by modern industry, which uses instead automated, horizontally laid-out factories employing cheap, unskilled labor and often based on truck transport. As an agglomeration of social fixed capital, New York is now obsolete and so, like an obsolete machine, is being abandoned by manufacturing firms.

One additional aspect of agglomeration theory is important: the location of several establishments belonging to a single firm. Suppose for example, that a firm wishes to set up one factory producing good X, another producing good Y, a research laboratory, and a headquarters facility. Should these all be in separate places or all in one place?

The firms's decision can be analyzed as follows. First, examine the transport and production costs of each facility in turn. We have already shown how to do this and discussed the structure of costs for each type of establishment. Second, the firm must decide what savings could be achieved if the facilities were combined. Do the two factories share services or transfer products? Would the research laboratory save by being adjacent to a factory (easy testing of ideas) or the headquarters (access to decision making and promotion of innovations)? Would the headquarters save by being near the laboratory or factory (cheaper day-to-day control and communication, but less focus on long range rather than immediate problems)? Figure 5.2 shows the savings that the firm thinks could be achieved by the headquarters, factories, and laboratory in one facility. Now, around the least cost sites of each single establishment are drawn contour lines of production and transport costs, showing how these costs rise from the least cost site. Thus at the $1 million isoline around X, transport and production would cost $1 million more than at X. Such contour lines are drawn around each site. The dark shaded area thus represents all those places at which the total additional costs of operating each facility away from its optimal location are less than the firm's anticipated savings from agglomeration. They are therefore all places at which the firm could profitably combine its facilities. (If there were no place at which the total additional costs of operation were less then the savings, than all facilities would disperse.) In Figure 5.2, the facilities should agglomerate.

Conclusion

Whereas the previous chapter focused on the effect of transport costs on location, this one has examined production costs and agglomeration economies. The most important spatially varying production cost is the cost of labor for industry as a whole (although particular industries may exhibit particular locational needs). These are supplemented by economies of agglomeration, notably produced and shared items of social fixed capital. We must reiterate the warnings given at the end of the previous chapter—these items (labor costs and social fixed capital) are themselves produced by the effect of earlier location decisions on the social environment. It remains to apply these ideas in practice, that is, to combine the theory of Chapters

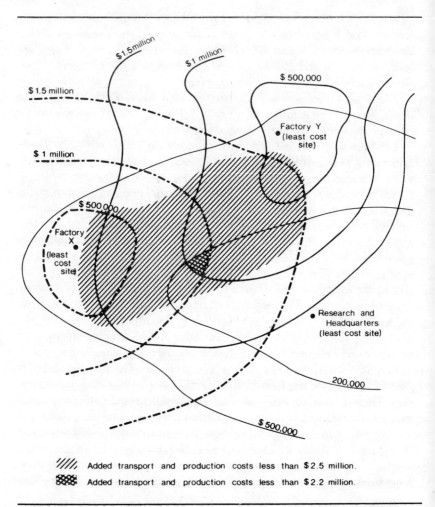

///// Added transport and production costs less than $2.5 million.
▧▧ Added transport and production costs less than $2.2 million.

Figure 5.2 Analysis of Agglomeration of Production Facilities of a Firm

3, 4, and 5 with the ideas on social organization in Chapter 2, in order to understand changes in actual location patterns.

Further Reading

Peter Lloyd and Peter Dicken's *Location in Space* (1977, pp. 197-299) contains a mass of data on production costs and agglomeration economies in Canada, the United Kingdom, and the United States. It is by far the best general souce of data; on more specific issues, you should follow up the references given in the text of the chapter.

6. INDUSTRIAL LOCATION IN PRACTICE

Location theory is one of the principal bases of geography. Agricultural location and land use theory is one of the principal supports of agricultural geography. Central place theory is crucial to urban geography; and industrial location theory underlies both urban geography and regional development as well as forming one source for spatial interaction theory.

This book has so far presented an introduction to the theory of industrial location. It has emphasized that location must be studied in the context of a given social and economic system. At present, in North America location can be interpreted as the interplay of two forces within the framework of least cost theory: One is the attempt to minimize transport costs, a factor of diminishing significance; the other is the effort to minimize labor costs. An important argument is that labor costs, like agglomeration economies, are not simply given but are themselves produced by the operation of the economic system in the past.

Now it is time to apply some of these ideas in practice—to see how the task of understanding some real-world location patterns and changes forces us to introduce additional elements to our theory. The following sections of this chapter contain examples of the location of the aircraft parts industry in New England, the spatial component of British industrial decline, and the distribution of industry within metropolitan areas.

Aircraft Parts Industry in New England

New England is one of the regions of the northeastern United States that shows most clearly the pattern of relative growth and relative decline that characterizes regional economies. The aircraft parts industry exemplifies this pattern; it shows also how the perceived locational attributes of the region for the industry have changed over time. Studies of the industry have been made by Estall (1966, pp. 157-170) and Bluestone et al. (1981); this section draws from these studies.

The pattern of location of the aircraft industry was established soon after World War I provided a secure market for aircraft, and the broad outlines of this pattern have changed little since then (except during periods of very rapid expansion, such as World War II). By 1972, nearly 95 percent of the value of complete commercial aircraft was produced on the West Coast (Washington and California), although military aircraft production was more dispersed (40 percent being in the South). Aircraft engines, however, were produced overwhelmingly in the Northeast and North Central states: Connecticut and Massachusetts alone produced 45 percent of all engine parts and accessories. The main producers in New England now are Pratt and

Whitney, General Electric (both making large jet turbines), Sikorsky (helicopters), Hamilton-Standard (propellers and parts), and Avco-Lycoming (small jet turbines). Supplying these producers are a variety of small and medium-sized subcontractors.

The production of airframes on the West Coast is resource-oriented—oriented, that is, toward locations where large factories do not have to be heated and where, for a high proportion of the time, the weather is good enough for flying. Good weather is vital for airplanes flown by visual rules. Once located, however, the huge investments in factories and production facilities effectively prevented the firms from relocating except under intense government pressure.

Just as airframe production became concentrated on the West Coast, so propeller and engine production was attracted to the Northeast. In this case, the important factors were the metalworking facilities and skilled work force of the Northeast. Few other areas could offer so many laborers skilled in metalworking or such metalworking facilities, particularly for products that had to be made accurately. Now, the large supply of engineers and scientists produced in New England helps keep engine production there. Indeed, some of the important aircraft engine producers evolved from other metalworking concerns: The Pratt and Whitney engine plant took over the facilities and skilled labor force of the earlier Pratt and Whitney Tool Company of Hartford, Connecticut. Hamilton-Standard, however, shifted to Hartford from Pittsburgh in 1931, after being bought by the same conglomerate (United Aircraft) that owned Pratt and Whitney—a move that centralized production and management in one area. Essentially, then, this interpretation of the engine and propeller segment of the aircraft industry concerns both labor (but in terms of availability and skill rather than wage) and agglomeration economies. These characteristics of New England were themselves produced by its earlier metalworking tradition. It is evident in the location of both aspects of the aircraft industry that transport costs have little effect.

Superficially, at least, the importance of New England in the aircraft industry has been increasing over recent years. For the aircraft industry as a whole, New England's share of the U.S. value of shipments was 13.0 percent in 1976 (a 25 percent rise since 1964) and its share of U.S. value added was 11.0 percent (up 8.4 percent since 1964). Hence the ratio of value added to value of shipments has fallen: Less of the final value of shipments is being manufactured in New England while a growing proportion is being made elsewhere and shipped to the region for final assembly. The region's aircraft industry is now less healthy than it appears at first sight: The advantages of New England as a location seem to have diminished over time. Two changes underlie this trend. On the one hand, the major producers—Pratt and Whitney and General Electric, for example—buy more parts from subcontractors than they make in-house: Over 50 percent of engine parts are bought from subcontractors. On the other hand, more and more of this subcontracting occurs outside the region than within it. There is little evidence that labor costs, transport or energy charges, or taxes are higher in New England than elsewhere in the United States, so why have these changes occurred?

One background factor must be mentioned first. In the theoretical discussion of least cost models, it was pointed out that as transport costs fell, so the attraction of cheap production cost sites became more significant. Undoubtedly, the dispersal of parts production from New England would have been more costly had it not been for the postwar reductions in long distance transport and communications costs.

But why do aircraft engine producers buy from subcontractors rather than produce in-house? Theory suggests that if a subcontractor could supply many firms, it could gain economies of scale in the production of components that were needed infrequently by any individual firm. But this does not apply in this case, for there are only two large jet engine producers in the United States and most subcontractors sell only to one firm. Two other reasons are more important. First, large firms use subcontractors to guarantee supply of an input. Secondary sources of inputs and parallel production by several subcontractors are used to reduce any disruption caused by work stoppages or other failures. Such backup facilities are especially important in the aircraft industry, where delays can mean the loss of most of a market (as, for example, the DC-10 dominates the L-1011 and the 707 dominated the slightly late DC-8). Second, subcontractors can be used to reduce the effect of sales fluctuations upon large firms. The demand for both military and commercial aircraft fluctuates widely, which implies that the use of the industry's labor and productive facilities should vary too, causing profit rates to fluctuate. Suppose that an industry employs 1000 people, 800 of whom make parts and 200 assemble these parts. If there is one firm and no subcontracting, a 50 percent fall in demand translates into a fall of 50 percent in that firm's sales. But suppose that there is one large firm, employing 400 producers of parts and 200 assemblers, and a set of subcontractors who together employ another 400 workers. Now a 50 percent fall in demand can be handled by the big firm cutting its work force to 300 producers and 100 assemblers (67 percent of the original work force) while buying in parts from 100 producers (to keep some of the subcontractor in the business). By subcontracting and varying its purchases from these subcontractors, the prime firms can partially isolate themselves from the effects of market fluctuations.

Finally, then, why does this rising level of subcontracting and pooled production translate into locational changes in the industry? Falling transport costs allow, but do not demand, dispersal of these plants from New England. Parallel production facilities have been set up by the two large engine producers in Quebec, West Virginia, and Vermont, which in contrast to their main facilities are nonunionized. Such plants provide production capacity during labor disputes and thus weaken workers' ability to strike. To do this, of course, the new plants must be geographically separated from the old ones. Subcontractors have been subject to a different set of pressures: The big firms employ geographically more dispersed subcontractors than before, partly to spread the risk of stoppages, and partly to avoid the problems created by recent shortages of supply of skilled labor in New England. It is also claimed that subcontractors in New England find it more difficult to borrow investment capital than do subcontractors elsewhere, but it is not clear why this is so.

This examination of the aircraft parts industry in New England has confirmed some aspects of the least cost theory in Chapters 4 and 5. The relative importance of labor and agglomeration economies has been emphasized in the context of an industry to which transport costs are of little significance. But the example has also introduced some novel features, especially the issues raised by the relations between big firms and their subcontractors. A more complex theory of location must evidently treat the questions of parallel production and multiple sources of inputs.

Industrial Decline in the United Kingdom

The recent decline of the manufacturing industries in the United Kingdom is well known. Between 1966 and 1981, the number of people employed in manufacturing in the United Kingdom fell from 8.6 million to 6.0 million (Massey and Meegan 1982, p. 5). The decline has not affected all industries and regions in the same ways; this section draws on the work of Massey and Meegan to examine the spatial implications of the decline. The most important theoretical point to be made by the example is that the geographic distribution of industrial employment must be understood in terms of the performance of the economy as a whole: In particular, the geography of job loss can be understood only by studying the causes of that job loss—the geography of employment decline cannot be understood simply in terms of the characteristics of different areas of the country. The example also shows the extent to which location theory can be used to analyze decisions other than the location of a single new plant.

Massey and Meegan studied 31 different industries, ranging from iron and steel through jute manufacture, from grain milling through scientific instruments. They concentrated on the period 1968-1973. Within these industries, three specific forms of the reorganization of industrial production were identified: (1) intensification, (2) investment and technical change, and (3) rationalization. Each has different implications for location patterns.

One form of production reorganizaiton that causes employment reorganization is *intensification.* This means the reorganization of a production process to increase productivity without closing plants or making major investment in new forms of production. In practice, then, intensification means fragmenting tasks, speeding up conveyor belts, new mechanization, or greater incentives. Intensification, then, is cheap, and so a favored way of raising productivity when profits are low (for instance during a depression). Then, if sales are stagnant, intensification implies job loss. Of the 31 industries, 6 followed this route: In the case of bicycle manufacture, for example, sales rose by 6.2 percent and productivity by 16 percent but employment fell by 9 percent over 1968-1973.

In 16 of the 31 industries, employment decline was accompanied by heavy net capital investment, usually in a major reorganization of the production process that resulted in much less labor per unit of output. This is *investment and techinical change.* The physical changes included mechanization, increases in scale, and alteration of the commodity. One example is the production of aluminum: A 25 percent increase in output was accompanied by a 40 percent increase in productivity, causing an 11 percent decline in employment. (Thus, large-scale capital investment and sales growth need not be associated with employment growth.) Generally this investment occurred in a period of slow growth or decline, and so it was not embodied in new, extra capacity but in replacement capacity.

The third and final form of production reorganization is *rationalization.* Unlike the other strategies, this produces job loss by direct disinvestment: complete or partial plant closure, scrapping of capital equipment, and cutbacks in the labor force. There is no major reinvestment in plants or machinery, the only alteration being one of scale. Generally this strategy is caused by lack of profitability: If an industry is unprofitable, capital is withdrawn to be reinvested in sectors with higher rates of profit. (This does not necessarily mean that the closed plant is unprofitable; only that it is less profitable than others.) Nine industries suffered job losses because of rationalization. In iron castings, for example, output fell 4 percent, productivity rose 19 percent, and employment fell 18 percent. (Presumably productivity rose as the least productive plants were closed.)

Each strategy has a different effect on employment patterns. Intensification reduces employment at existing plants, involving neither plant closure nor new investment. Its effect is confined to the existing plant sites. Unequal regional effects are produced not by job transfer or job creation, but by the differential distribution of job loss. Rationalization also involves no new locations, so changes in employment all occur within the existing geographic distribution. It may, however, mean employment growth at some sites (e.g., large, centralized factories if smaller outlying plants are closed down). Some regions may gain a small number of jobs at the expense of other regions' loss of many jobs. Rationalization may imply the complete closure of some plants. By contrast to the first two strageties, technical change always involves net investment, so there is always some decision to be made about the location of that investment. In periods of stable or falling sales, this new investment is accompanied by plant closures elsewhere.

This argument may be summarized as follows:

Intensification:	employment losses at the site;
Rationalization:	employment losses at the site, plant closures, plus possibly some small local gains as capacity is centralized;
Technical change:	employment cuts at the site, or new locations and closures.

An example of rationalization has been chosen for more detailed analysis: the iron castings (foundry) industry. Rationalization is the reverse of the location decision, concerning which factories or parts of factories to close. Our theoretical criterion is that the least profitable capacity is scrapped first. Least cost theory identifies attributes of location that affect profitability—especially, in this example, labor costs (and militancy) and accessibility to growing markets. To the analysis of this theory, however, must be added the characteristics of the plants themselves: labor productivity and economies of scale, for example. Clearly locational factors will have geographical effects (some regions have more jobs than others); but it does not follow that regional variations in rates of job loss must be due to the locational factors, for productivity and plant size may themselves vary regionally.

Nearly 16,000 jobs were lost in the iron castings industry between 1968 and 1973. Figure 6.1 shows the geographic distribution of these losses, which varied from 11 percent in the Midlands to 35 percent in Wales. Variations in the losses are attributable to two factors. First is the degree of output change in each region as a consequence of specialization in supplying different industries. Output rose in the South, and was more or less stable in Yorkshire, the Midlands, and Wales; the other regions, in which output fell, all lost over 20 percent of their employment in this industry. This, then, is a geographic location factor—namely, access to growing markets. The second factor concerns plant characteristics. There was a far higher closure rate among small foundries than among large, and an associated tendency for productivity increases to occur more quickly in some regions than others.

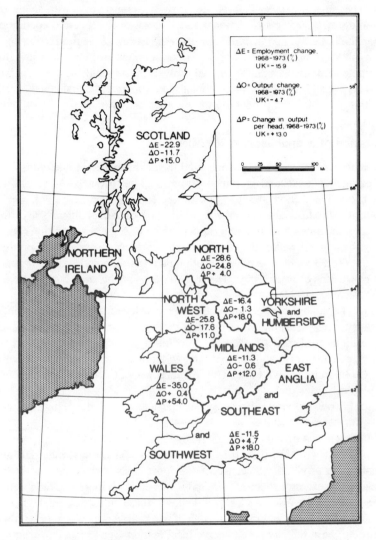

SOURCE: Based on Massey and Meegan 1982, pp. 150-151.

Figure 6.1 Regional Employment, Output and Productivity Changes in the UK Iron Castings Industry, 1968-1973

Two lessons should be drawn from this example. The first is Massey and Meegan's insistence that locational analysis must begin with an understanding of the effects of national and international economic conditions upon the industry, and in particular upon the nature of the production decisions that are being made. Second, in the context of plant closures, this example demonstrates that locational analysis must include not only the factors discussed in least cost theory but also factors internal to the plant (such as its age, scale, and productivity). Another extension to the theory must be made.

Location of Manufacturing in Cities

Whereas the first example analyzed the performance of one industry in a region and the second examined the regional performances of several industries, this section uses location theory to understand the evolution of patterns of industrial location within U.S. and Canadian cities. The example demonstrates how the general principles of least cost location theory must be applied within the context of changing characteristics of production in order to understand how location patterns change: Location depends on production characteristics. The section draws on the work of Scott (1982) and Webber (1983).

We begin by considering manufacturing location within nineteenth-century cities. Several characteristics of these cities were important. (1) Near the central core were located the intercity transport terminals (ocean, river, canal, or rail) and the surrounding labor force, living at high density. (2) Intercity freight movement was far cheaper than intercity transport. (3) Within cities, people could be moved more cheaply than could freight. Also, (4) raw material processing was inefficient, so that much weight was lost during manufacture.

Four classes of industry were recognized in nineteenth-century cities. The first consisted of industries directly tied to on-site raw materials and small-scale production because of transport costs: breweries and brickmakers, for example. Second were industries in which there was little weight loss during processing. As Chapter 4 shows, these industries sought market locations—either at intercity terminals, if markets were nonlocal (e.g., shipbuilding) or at the city core to serve local markets (newspapers, furniture, baking). The third group of industries suffered large weight losses during processing and so had to be located near the source of raw materials—that is, the transport terminals. The waterfronts of Toronto and Montreal still exhibit some relics of these industries—sugar-refining and flour-milling. mainly. Fourth were the industries that need little materials but much labor: Being small scale and labor intensive, such industries as clothing, printing, and precision manufacturing located near the source of labor, the city core.

Thus we understand why, with labor, intercity and local markets, and raw materials all located near the city cores, manufacturing in nineteenth-century cities was so highly centralized.

During the twentieth century, however, the concentrated, high density nineteenth-century manufacturing city has exploded over the surrounding countryside. Throughout North America, the inner cities have been losing manufacturing jobs whereas the outer areas of metropolitan regions have been gaining. The problem, then, is to determine how these locational changes are related to the evolution of the economy.

Several processes have been occurring (see also Chapter 2). One is competition between firms, which leads them to try to reduce work by mechanization and increasing plant size. Accompanying this technical change has been a search for new markets. As industrial markets become saturated (remember the product cycle hypothesis), new products and new markets must be found. Both processes reflect an ever-increasing emphasis upon innovation and have been accompanied by a significant lowering of transport rates (at least until recently, when world oil price rises have affected transport costs). These processes have had three important manifestations. (1) Increases in plant size have been accompanied by increasing concentration of market power in a few firms in each industry (although some industries are still characterized by small firms). This, in turn, has had several consequences. Plants are now less likely than they were to serve a single city, and so the attraction of the CBD as the point of easiest access to the city market has diminished. In addition, conglomerate ownership of several plants appears to have made locational change easier and faster (Massey and Meegan, 1982, discuss this issue). As plants have grown larger, transport costs have fallen, so access to terminals and to labor has become less significant than spatial variations in production costs—especially the cost of land. (2) New commodities have been continuously introduced by firms to replace goods and services previously produced by the domestic economy, so now there are many more industries than there were. (3) The state has provided several new markets for firms to exploit. In the United States, the arms and space industry is one of them. But more important has been the state's attempt to increase (and stabilize) demand by establishing high levels of investment in housing, urban freeways, transit systems, and automobiles.

The locational effect of these processes upon industry in cities has been dramatic. (Still, the small-scale, labor-intensive, and the site-oriented industries have been less affected than the others. Typically, these industries remain downtown. But because of product innovation, these industries are of declining importance.) Intraurban freight rates have fallen and materials processing has become more efficient. Technical changes have replaced labor by machinery and land. The work force has been suburbanized. Also, the CBD has become highly congested. Thus, as the materials-intensive industries (the second and third types) were freed from their ties to central terminals and as the labor force and local markets dispersed to the suburbs, so needs for space and CBD congestion pushed them away from central locations too.

This example demonstrates, then, how the changing balance of transport and production costs at different locations has altered the geography of manufacturing industry within cities. In turn, however, changes in production and transport costs must themselves be understood as a result of mechanization, scale changes, new products, and state provision of markets. Again, then, we see how economic changes lead to locational changes.

Conclusion

This book has provided only an introduction to the theory of industrial location. Nevertheless I have tried to present a modern view of location theory, one that introduces you at least to some current research problems. Thus the "introductoriness" arises from simplification of issues, evasion of some problems with the argument, avoidance of formal (and often, mathematical) treatments of location theory and a failure to treat adequately such issues as industry structure and the internationalization of production. Yet the main arguments are not "introductory" ones: The notion of maximum profit and least cost location is central to location theory; around this idea must be stated several issues of method—the importance of historical context, the need to understand production decisions as a prelude to location decisions, the falling significance of transport costs, the prominence of labor costs and agglomeration economies among production costs, and the fact that labor characteristics and the agglomeration economies at places are both produced characteristics. The examples have shown how these themes must be elaborated within a more complete theory of location; this book should be considered a first,not the last, word on location theory.

You will learn much more about industrial location if you read four additional books. The first is Lloyd and Dicken's *Location in Space*. This is a traditional but very good treatment of both theoretical and empirical aspects of location. Second, consult the two studies upon which the first two sections of this chapter are based: Bluestone et al.'s examination of the aircraft industry in New England, and Massey and Meegan's *Anatomy of Job Loss*. These will take you far into ideas of context, labor costs, and industrial structure. Finally, Rees, Hewings, and Stafford have published a collection of papers, *Industrial Location and Regional Systems,* which will give you an idea of the scope of modern industrial geography.

REFERENCES

Alchian, A. A. 1950. Uncertainty, evolution and economic theory. *Journal of Political Economy* 58: 211-221.

Baumol, W. J. 1959. *Business behavior, value and growth.* New York: Macmillan.

Berry, B.J.L. 1967. *Geography of market centres and retail distribution.* Englewood Cliffs, NJ: Prentice-Hall.

Bloom, C. C. 1955. *State and local tax differentials.* Iowa City: State University of Iowa.

Bluestone, B., Jordon, P., and Sullivan, M. 1981. *Aircraft industry dynamics.* Boston: Auburn House.

Cyert, R. M., and Marsh, J. G. 1963. *A behavioral theory of the firm.* Englewood Cliffs, NJ: Prentice-Hall.

Estall, R. C. 1966. *New England: A study of industrial adjustment.* London: Bell.

Estall, R. C., and Buchanan, R. O. 1973. *Industrial activity and economic geography.* 3rd ed. London: Hutchinson.

Haggett, P., Cliff, A. D., and Frey, A. 1977. *Location models.* London: Arnold.

Hamilton, F.E.I., and Linge, G.J.R. 1979. *Spatial analysis, industry and the industrial environment: 1, Industrial systems.* New York: John Wiley.

Harris, C. C., and Hopkins, F. E. 1972. *Locational analysis.* Lexington, MA: Lexington Books.

Harrison, B., and Kanter, S. 1978. The political economy of states' job-creation business incentives. *Journal of the American Institute of Planners* 44: 424-435.

Haynes, K., and Fotheringham, A. S. 1984. *Gravity and spatial interaction models.* Beverly Hills, CA: Sage.

King, L. J. 1984. *Central place theory.* Beverly Hills, CA: Sage.

Lloyd, P. E., and Dicken, P. 1977. *Location in space.* 2nd ed. New York: Harper & Row.

Lowe, J. C., and Moryadas, S. 1975. *The geography of movement.* Boston: Houghton Mifflin.

Massey, D. 1979. A critical evaluation of industrial-location theory. In *Spatial analysis, industry and the industrial environment: 1, Industrial systems,* eds. F.E.I. Hamilton and G.J.R. Linge, pp. 57-72. New York: John Wiley.

———and Meegan, R. 1982. *The anatomy of job loss.* London: Methuen.

Phillips, E. B., and LeGates, R. T. 1981. *City lights.* New York: Oxford University Press.

Rees, J., Hewings, G.J.D. and Stafford, H. A. 1981. *Industrial locations and regional systems.* New York: Bergin.

Scott, A. J. 1982. Location patterns and dynamics of industrial activity in the modern metropolis. *Urban Studies.* 19: 111-142.

Smith, D. M. 1979. Modelling industrial location: Towards a broader view of the space economy. In *Spatial analysis, industry and the industrial environment: 1, Industrial systems,* eds. F.E.I. Hamilton and G.J.R. Longe, pp. 57-72. New York: John Wiley.

94

Statistics Canada. Various years. *Census of population*. Ottawa: Author

———Annual. *Manufacturing industries of Canada: national and provincial areas* (catalogue 31-203). Ottawa: Author.

———Annual. *Manufacturing industries of Canada: subprovincial areas* (catalogue 31-209). Ottawa: Author.

——— 1972. *Standard industrial classification manual* (revised 1970). Ottawa: Author.

Thompson, W. R., and Mattila, J. 1959. *State industrial development*. Detroit: Wayne State University Press.

Treysz, G., Friedlaender, A., and Tresch, R. 1976. An overview of a quarterly econometric model of Massachusetts and its fiscal structure. *New England Journal of Business and Economics* 1: 57-72.

U.S. Department of Commerce, Bureau of the Census. Series. *Census of manufactures*. Washington, DC: U.S. Government Printing Office.

———. *County business patterns*. Washington, DC: U.S. Government Printing Office.

———. *Historical statistics of the United States, colonial times to 1970*. Washington, DC: U.S. Government Printing Office.

———. 1981. *Statistical abstract of the United States*. Washington, DC: U.S. Government Printing Office.

Webber, M. J. 1983. Location of manufacturing activity in cities. *Urban Geography* 3: 203-221.

Weber, A. 1909. *Theory of the location of industries*. Chicago: University of Chicago Press.

Yeates, M. 1975. *Main street*. New York: Macmillan.

ABOUT THE AUTHOR

MICHAEL J. WEBBER is Professor of Geography at McMaster University in Hamilton, Ontario. He received his Ph.D. from the Australian National University in 1967. Professor Webber is the editor of *Papers, Regional Science Association* and on the editorial boards of *Geographical Analysis* and *Journal of Regional Science.* He has published several books on location theory—*Impact of Uncertainty on Location* and (coauthored) *Christaller Central Place Structures*—and on operational urban models—*Information Theory and Urban Spatial Structure* and *Explanation, Prediction and Planning.* His papers have appeared in a variety of geographical journals. Professor Webber's research interests now include both industrial location theory and regional economic development, particularly in relation to technical change and economic cycles.